Nuclear Structure

Nuclear Structure

Rangel Kolessin

ISBN 978-0-557-19504-6

PREFACE

The importance of the structure of the atomic nucleus for science is significant. The fact that the matter consists of atoms and that each atom has a nucleus has been taken for granted by many books and articles. In all publications it is admitted that the problem of the nuclear structure is far from being solved. The present book proposes a new, adequate solution to this problem.

The scientific traditions require that every next generation follow the path of its predecessors. Without a doubt it is the beaten path, which is supposed to ensure success, but what to do when the path, although beaten, leads to nowhere? Obviously, a change in direction is needed. Instead of inventing a model for each nuclear property (and neglecting the other properties), it is necessary to look for one more general solution. The question is: *What would a nuclear structure, which is in full agreement with the laws of Physics and with all nuclear properties look like?* As far as I know the structure of DNA was discovered under similar conditions. May be the nuclear structure is even harder to discover due to the uncertainty about the structure of the nucleons.

The purpose of this book is to present *the real structure of atomic nucleus*, a structure which explains all nuclear properties without any additional assumptions. The proposed nuclear structure is *completely new* and has nothing in common with the existing nuclear models. As the goal of each model is to help reveal the nuclear structure, *the nuclear structure discovery will make all models pointless*. Therefore the list of references is very limited. There is no sense in prioritizing useless solutions. The main source of data used for the properties of the nuclei is the "Reference book of Nuclear Physics", O.F.Nemetz and U.V. Hofman [1]. The general interpretation of the problem is made according to the available reference books of Physics [2,3,4]. The history of the problem is stated according to S. Glasstone [5]. The latest state of the problem is in compliance with the "Proceedings of 8-th international spring seminar on Nuclear Physics" [6] and discussions on the Internet. I believe, in every country there are books on Nuclear Physics, as in Bulgaria for instance are the books of Slavov [7] and Balabanov [8], and those are available for all who need more information on the problem.

Because of technical difficulties, the numbers of protons and nucleons are shown after the symbol of the nuclides, for instance: C_6^{12}, instead of in front: $_6^{12}C$.

The book is intended mainly for physicists and experts in the area of Nuclear Physics, but it could also be of interest to all who love Physics.

CONTENTS

INTRODUCTION

The discovery of the atomic nucleus by Rutherford (1911) was one of the most important and most exciting events in Physics because it was the first step towards revealing the mystery of matter. Although the properties of substances depend on the electron structure of the atoms (and namely on their valent electrons), the structure of the nucleus also plays a very important role because the structure of the nuclear electric field determines the number and the arrangement of the electrons. But above all is the interest in nuclear energy - the hope and the fear of humanity. Thus revealing the nuclear structure is not only a scientific quest, it has implications for mankind.

As with most great discoveries, Rutherford's idea was at first rejected because the circulation of electrons around the nucleus, like any other motion, could not be endless. Later, in 1913, Bohr's theory and Moseley's experiments cleared up the doubts. In the micro-world perpetual motion does exist, although still no one knows why. The quest to discover the nuclear composition continued during the following 20 years until Chadwick's discovery of the neutron in 1932. Then the problem of the dependence of observed properties on composition arose, the answer to which was hidden in the nuclear structure. With the discovery of nuclear energy, Nuclear Physics was elevated to the level of leading science.

A century of hard work of several generations of scientists and their attempts to uncover the structure of the nucleus and to master the nuclear power have resulted in the following:

1. Tremendous amount of experimental data concerning all aspects of the nuclear structure and properties;

2. Very impressive achievements in harnessing the nuclear energy;

3. Very humble theoretical results in understanding the essence of the nuclear structure, connected with a lot of contradictory models. Although each model is only a working hypothesis, the contradictions between them, and the discrepancy between all models and the laws of Physics, make a strong negative impression. The absence of a single adequate theory, developed on the grounds of the general physical principles, is perhaps one of the most significant failures of science to date.

What has happened? A lack of eagerness or a deficit of talent? Certainly neither, because in their attempts to find a solution, scientists have shown incredible ingeniousness (honored by Nobel prizes in 1963 and 1975). For instance, scientists have found a reconciliation of the contradictory philosophy of the amorphous structure (*the liquid drop model*, developed lately in *ab initio models*) with an atom-like structure (*the shell model*) in a hybrid (*the collective model*). They have prescribed to part of the nucleons to form *a liquid drop core* and to create *a potential field* (self coordinated mean field), necessary for the other part of the nucleons to move in *certain orbits* like the electrons in atoms. *Pauli's exclusion principle* determines which of the nucleons must be in the core, which in the orbits (*valent nucleons*). The tightly tied *in pairs* protons and neutrons look like *quasi-particles* and create (in a fantastic way) superconductivity and superfluidity in the nucleus. Nevertheless, each ingenious answer raises even more questions and becomes the ground for a new stage of work.

Thus, the main reason behind this eluding solution is one: *Difficulties!* The enormous torrent of difficulties follows directly from the very definition of the nucleus. *The atomic nucleus is a quantum system, which consists of many strongly interacting particles.* This definition was accepted a priori (by scientific intuition) and so, a priori, a torrent of unexpected difficulties arose. It is well known, that the problem of interaction of only 3 bodies, which attract each other with well defined force (for instance a force inverse proportional to square of the distance), has not yet full theoretical solution. In a nucleus the number of the bodies (nucleons) can approach 300 and the attractive force between them (the force of the strong interaction) is still unknown. In addition, the system of nucleons in a nucleus is quantum one. This presupposes solution of Schrodinger's many particles wavy equation with unknown Hamiltonian. The nucleons are fermions, i.e. they have spin ½ η and obey the statistics of Fermi-Dirac. The nucleons possess magnetic moment – different by values and sign for protons and neutrons. Hence, the strong interaction must be in agreement with the unknown way of spin and magnetic moment compensation (because the spin and magnetic moment of the nucleus are zero or have values near to the values of one nucleon). Except that, the protons have electric charge. The absence of nucleus, which consists only of protons, imposes the necessity of correlation between the electrostatic repulsion and the unknown force of the strong interaction. Finally, there is an unknown repulsive force between the neutrons in the nucleus, which explains the limited number of neutrons in a nucleus.

How to find solution to such a difficult, theoretically insoluble problem? The intuition suggests: "step by step", using all possible approaches, creating different models, searching analogy in other branches of physics, looking for all possible correlations, believing that one day the quantity will turn into quality. Several generation of scientists have worked according to this plan. The whole arsenal of Mathematical physics (symmetry, transformation, algebra, theory of groups, tensor operators etc.) is set to work for model creation. As a result enormous mathematical models have been created. The goal of each model is to explain a specific nuclear property, and although they used different mathematical tools, each model has in its base one of the physical nuclear prototypes – liquid drop, shell model, independent particles etc. Each nuclear property has been modeled by different approaches, using different mathematical tools, imposing different interpretations. The biggest part of these models is devoted to nuclear spectrum, because of the belief that the disposition of the energy levels ought to be connected by the nuclear structure, and because of the abundance of experimental data. Despite the "genetic" tie between the physical origin and mathematical "image", correlating the parameters of mathematical models and nuclear properties is a difficult task. The work of modeling continues with accelerated rates.

This system of work has several very serious defects:

1. An analysis shows that all nuclear properties depend mainly on the number of nucleons- on the sort (even or odd) of these numbers, on the proton/neutron ratio, on the "magic" of the numbers, etc. In view of the charge independence of nuclear forces, the role of the sort of nucleons and the sort of numbers rest inexplicable. This testifies that many bodies interaction cannot be the basic factor in nuclear structure. The nuclear structure is first of all the *way of nucleon arrangement*. The binding energy (which should be a direct product of the "many body interaction") depends mainly on the nucleon arrangement.

2. The physical nuclear prototypes, used as basis for the mathematical models, contradict many physical phenomena, as for instance the electrostatic repulsion between protons (thesis about the charge independence of nuclear forces), free

propogation of electrostatic field and its shielding by neutrons. The problems of nucleon spin and magnetic moment compensation is simply ignored. Evidently *the role of protons and neutrons in nuclear structure* is not completely realized.

3. The neglectful attitude towards the nucleon structure is unacceptable. Clearly, the nucleons cannot be structureless particles, because each one of them has its own properties. It is the structure which determines the properties. *The particle does not consist of mass, charge, spin and magnetic moment, but each of these properties correspond to a specific structural element of the particle.* The interaction between particles is not an interaction of the properties. The particles interact by means of their structural elements, which leads to changes in the properties.

The *form and the size* of the nucleons are of exceptional importance for the structure of a nucleus. The scientific intuition suggests as a most probable a *spherical,* point-like form of the nucleons. Such form suggests *the spherical* (or a sort of spherical, volumetric) *form of the nucleus as the only possible one.* Perhaps this form appears very natural, but *the arrangement of spherical strong-interacting nucleons in a spherical nucleus is not a monovariant one!* If the forces are strong enough to keep the nucleons firmly together, nuclei with different structure and different properties will be obtained for any given numbers of protons and neutrons. But nuclear isomorphism does not exist! If the forces are not strong enough and the nucleons could move freely, the nucleus would be structureless (*chaotic structure*) and the nuclear properties would be undetermined.

So, should the nuclei creation depend on the science, it would be impossible. The Universe would not exist. But since the existence of nuclei is out of any doubt, obviously Nature has managed to avoid all scientific difficulties. Nature's acts are not based on intuition! Nature has created nucleons in a form suitable for formation of the nucleus and has been foreseen in nucleons structural elements appropriate for the forces of strong interaction. Thus we are obliged to follow the natural way – *analysis instead of intuition.*

A special note to the US reader:

The author uses the European system for **decimal and thousands separation:**
For example, he uses decimal *comma* instead of decimal point.

Also, when indicating multiplication, he uses a point, instead of the "X" symbol.

CHAPTER 1

PRELIMINARY ANALYSES

1.1 CLEUS STRUCTURE ANALYSIS

Let us assume temporarily the idea of a spherical form of the nucleons and try to construct a spherical nucleus, observing all laws of Physics, formulated as preconditions:

- *The electrostatics precondition.* It is reasonable to assume that the spherical protons form electrostatic field with a spherical symmetry - in view of the spherical symmetry of the electron configuration in the Hydrogen atoms (recall space quantization). The electrostatic field of protons must propagate freely up to where the electrons are. Although the electromagnetic interaction is considered to be approximately 10 times weaker than the strong interaction [2, 3], evidently at very sort distances (direct contact) the repulsive electrostatic force becomes capable of overcoming the attraction force of the strong interaction. There is no other sensible explanation of the absence of "diprotonium" (He_2^2)! Hence we do not have justification for putting two or more protons side by side in a stable nucleus. The neutrons in the nucleus must simultaneously play two roles - *to shield the electrostatic fields and to bind the protons together*. The belief in the charge independence of the strong interaction must be put aside until its explanation can be derived from the structure of the nucleus. When constructing the spherical nucleus we must keep in mind that *a free propagation of the electrostatic field is incompatible with shielding!*

- *The "mechanics" precondition*: Both types of nucleons (protons and neutrons) have equal spins (J= ½η). Although the spin is considered an "intrinsic" angular momentum, the formula $J = mcr$ shows that the mass of the nucleon (m) rotates with the speed of light (c). The radius (r) need not necessarily be the radius of the outside surface of nucleons (if such surface exists). Irrespective of its value, the radius of rotation exists and a real angular momentum also exists. Therefore, we can not afford to ignore the presence of nucleon rotation. But, when constructing the spherical nucleus from rotating nucleons, we must keep in mind the condition that strong ties bound the nucleons with their neighbors – a condition incompatible with free rotation.

Nuclei made out of spherical nucleons must possess at least the following properties:

1. The electrostatic field of a nucleus must be equal to the sum of the electrostatic fields of all protons. No shielding is permitted.

2. The spin of a nucleus in the ground state is zero or is equal to one or several values of nucleon spin;

3. The magnetic moment of a nucleus is either zero or lies between zero and several values of a nucleon magnetic moment;

4. The nuclei, which have unpaired (or extra) nucleons, possess a quadruple moment – a static characteristic, incompatible with dynamic structures (liquid drop, core, orbits)!

5. Generally the spin and the magnetic moment of the nuclei do not depend on the number of nucleons;

6. Very strong ties exist between at least several neighboring nucleons, imposed by the forces of strong interaction.

First, it becomes clear that the above-mentioned properties of the nuclei cannot be obtained by a random distribution of nucleons. But random distribution is the basic property of any liquid drop or liquid core. Many properties of the nuclei, such as spin, magnetic moment, quadruple moment, testify to a strict nucleon arrangement. However, a strict arrangement is practically incompatible with nucleon motion. Next, it becomes clear, that there is *neither theoretical nor practical possibility* for spherical nucleon arrangement in a spherical nuclear structure, which could provide the necessary nuclear properties! It is not possible that a *spherical or any other volumetric form of a nucleus can be made out of spherical nucleons!* Although it is quite obvious, here is a short proof.

Fig.1 shows a schematic cross-section of a spherical middle mass nucleus obtained by an arrangement of spherical nucleons. The number of nucleons predetermines the existence of two coordination spheres (of nucleons) around a central neutron. The protons are marked by small crosses (plus signs) as a symbol for positive charge.

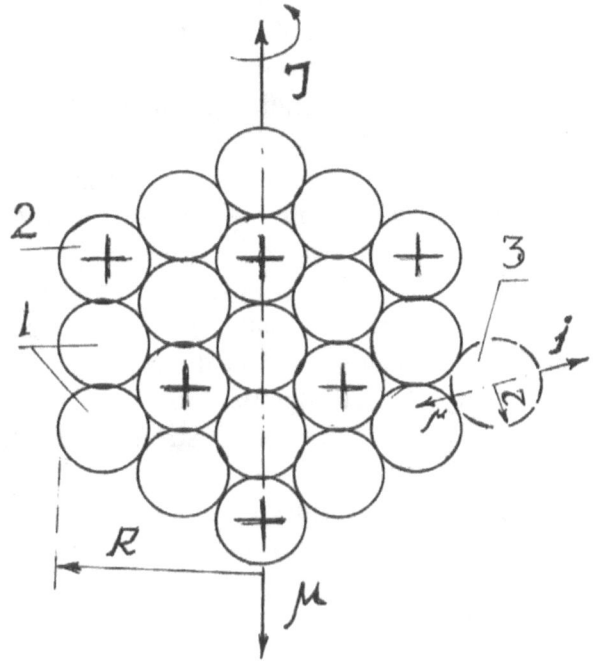

Fig. 1. Schematic cross-section of a spherical nucleus
1. Neutrons; 2. Protons; 3. Unpaired (extra) nucleon

From Fig.1 the following points and questions arise:
- The number of protons is strongly restricted by the necessity of neutron shielding. This leads to a neutron/proton ratio of approximately 3:1 – too high for a stable nucleus. It is well known, that even in the heaviest stable nuclei this ratio is in the range of approximately 1,5:1. For instance, if a nucleus with a full second coordination sphere contains approx. 64 nucleon – 16 protons and 48 neutrons, it ought to correspond to S_{16}^{64}. Of course, such nuclide cannot exist, because the neutrons "dripping" must occur at S_{16}^{52}. On the other hand, if the stable nuclide S_{16}^{32} have a spherical form each of the protons must contact with at least 5 other protons. But if such contacts are admissible the presence of neutrons in the nucleus would be useless.
- The electrostatic field of the protons in the inner coordination spheres is completely shielded by the neutrons from the outer sphere.

So in a middle mass nucleus more than half of the proton's electrostatic fields cannot get outside of the nucleus and thus, the number of the electrons in the atom must be considerably lower then the number of the protons! The shielding effect of "valent neutrons" (if such exist) also cannot be ignored. Any interruption of an electrostatic field will cause an immediate tearing of electrons from the atoms. Of course, an arrangement where all protons are on the surface of the nucleus must be rejected in advance because the presence of neutrons inside the nucleus would become unnecessary. Hence, the contradiction between the necessity of shielding and the free propagation of the electrostatic field of the protons is a reason enough for rejecting any idea of arrangement of spherical nucleons in a spherical nucleus. This conclusion is strong enough not to merit further discussion.

- If each proton is connected with a neutron (forming a proton/neutron pair), what would the role of non pared (excess) neutrons be? Aneutron segregation is forbidden (because even a "dineutronium" does not exist).

- The nuclear spin predetermines a rotation of the nucleus as a whole (rotation of a rigid body). As any angular momentum, the nuclear spin must depend on the nuclear momentum of inertia (i.e., it must depend on the mass and radius of the nucleus, or it must depend on the number of nucleons). Then, how could the nuclear spin be determined by an extra nucleon, like the one at the right side on Fig.1, which has an arbitrarily oriented spin? What if the projection of its spin on the axis of nucleus rotation is zero? Moreover, if the plane of rotation of this nucleon is parallel to the axis of the rotation of the nucleus, a Coriolis force would act and destroy the orientation. Even if the spin of this nucleon is parallel to the axis of the nucleus rotation, how should it influence the rotation of the nucleus?

- How does the compensation of nucleon spins with arbitrarily orientation occur? How does the compensation of nucleon magnetic moments occur at arbitrarily oriented nucleon spins? If the nucleus rotates as a whole, the proton charges will produce a considerable magnetic moment. But in all even/even nuclei the spins and magnetic moments are exactly zero. Thus it follows that the geometric sum of the projection of spins and magnetic moments of arbitrarily oriented nucleons cannot determine the spin and magnetic moment of a spherical nucleus.

- If the binding force is really a strong one, it will tightly press the nucleons against each other. The force indeed is very strong, because approximately 1% of the nucleon mass is spent as binding energy. What will happen to the separate rotations of tightly bonded bodies? Surely such a tight bond will stop any separate rotation. And if in some miraculous way the bond will not stop the individual rotation, the friction between the surfaces with opposite rotations will stop it for sure. As it is shown in Fig.2, a rotation would only be possible between two nucleons having anti-parallel spins. The addition of a third (or more) tightly bonded nucleon will stop any rotation.

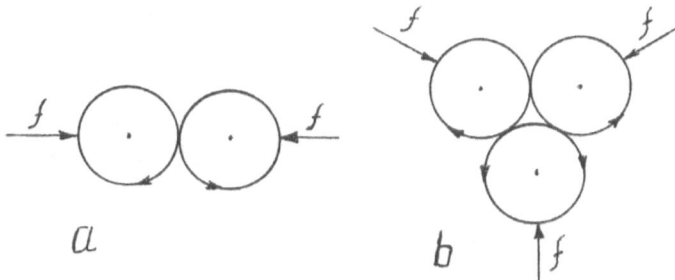

Fig. 2. Scheme of possible (a) and impossible (b) nucleon rotation

One of the most striking examples of the unsoundness of the idea of spherical form of the nuclei is the existence of the mesoatoms. It is well-known that mesoatoms have muon (μ^-), meson (π^-, K^-) or hyperion (Σ^-) in orbit instead of electron. The radius of the orbit r_o is inversely proportional to the product of the meson mass (m_m) and the number (Z) of protons in the nucleus:

$$r_o = 5{,}3.10^{-9} \frac{m_e}{m_m . Z} \text{ [cm]}$$

Where m_e is the mass of the electron.

Therefore, in middle and heavy mass nuclides, the radii of mesoorbit become less than the radius of the spherical nuclide (r_n) calculated from the well-known formula for the liquid drop model:

$$r_n = 1{,}3.10^{-13} \sqrt[3]{A} \text{ [cm]}$$

As an example, the radius of the nuclide Sn_{50}^{120} is 6,4 .10^{-13} cm, while the radii of its mesoorbits are: 5,11. 10^{-13} cm; 1,09. 10^{-13} cm and 4,5 .10^{-14} cm for orbits of μ^-, K^-, Σ^-, respectively. Hence, the orbits of all types of mesons should be located inside the nucleus of Sn_{50}^{120}. How about the heaviest mesoatoms? The radius of Σ^- in U_{92}^{238} is only 2,5x 10^{-14} cm. Of course, the explanations such as "mesons move through and inside the nucleus" should be rejected as very naïve. The experience with nuclear reactions shows that the nucleus does not bear any intervention from outside without a prompt reaction. A crucial question cannot be avoided. What really exists – mesoatoms or spherical nuclei? As the existence of mesoatoms is a fact, it becomes clear that the radius of the nucleus must be much smaller. This means that the form of the nucleus cannot be a spherical one.

The list of disproof is not exhausted and each example alone is enough to lead to the explicit conclusion that: *A spherical nucleus cannot be built up of spherical nucleons without violation of basic physical principles.* Hence clarifying the nuclear structure needs to start with a change of the basic concepts about the form and structure of the nucleons.

1.2. REMARKS ON NUCLEAR FORCES

Usually, the notion "nuclear force" is applied to the attraction force, which keeps the nucleons together, also known as a strong force or a force of strong interaction. But, as a matter of fact, all forces, which act within the nucleus, are nuclear forces. Without any doubt the electrostatic repulsive forces act within the nucleus. Of course, the strong attraction force is decisively the one responsible for the nucleus formation, but at very short distances the electrostatic force of proton repulsion becomes similar in magnitude to it, or even bigger. Perhaps this conclusion seems doubtful, but the absence of diprotonium (He_2^2) cannot be explained otherwise. Also, at very short distances between the neutrons a repulsive force of electrostatic origin (from residual electric fields) arises, which can reduce the attraction force and this explains the limited number of neutrons in the stable nuclei. The equilibrium between all nuclear forces is the basis for the stability of the nucleus.

Since the energy of the strong interaction per nucleon is equal to approximately 1% of the mass of the nucleons, the idea of massless gluons as binding agents is improbable. The exchange of pions as a way of binding is also improbable – there is no reasonable explanation how an exchange of spherical particles can produce an attraction force. The exchange of particles is an exchange of impulses and can only produce repulsion. Evidently, the attraction force arises from the process of mass transformation into energy – a process similar to the wave interference or matter annihilation. At any rate, the strong force needs a relatively large contact between nucleons. The spherical form of the nucleons is not appropriate for such contacts. Hence, the form, structure and nature of the nucleons ought to be in accordance with these requirements of the strong force.

1.3. REMARKS ON NUCLEON STRUCTURE

A definite object cannot be build from an indefinite material. So the form and structure of nucleons must be clarified in advance. Let us begin with rejecting some stereotypes and preconceptions.

The first preconception concerns the form of the nucleons. Where does the conviction that the nucleon has a spherical form come from? It might have resulted from an incorrect intuition: if the nucleons are considered as material points, and the points look like spheres, then the nucleons must be spherical… But if the nucleus cannot be built of spherical nucleons, why should we keep insisting that they were spherical?

The second preconception concerns "the inherent properties" of nucleons, like spin, magnetic moment, etc. Generally, the notion "properties" marks the fact of existence of a certain phenomenon, without any attention to its structure. In the case of complex objects this limited attention could be justifiable; however, in the case of the elementary particles each property is actually determined by an element of the structure. So, if we want to understand the structure, we are obliged to know the nature of each property.

On the other hand, in view of the existence of spin (rotation at the speed of light) the spherical form of nucleons is practically impossible. In case of a rotation at a speed substantially bellow the speed of light, the centrifugal force will give to the rotating body the shape of a disk or a washer. Then, why shouldn't we accept the disk-like form of the nucleon as the most probable? Surprisingly, the disk-like form turns out to be very promising. For each "inherent property" a corresponding structural element can be found. The convenient place for the electric charge is the rim of the disk, from where the electrostatic field spreads mainly in radial direction, at some angle on both sides from this direction. So the electrostatic field is laterally restricted by two conical surfaces, as it can be seen from Fig. 3 -a. This form of the electrostatic field fits the law of the inverse proportionality to the square of the distance. The spin will be the angular momentum of disk rotation around its axis. The magnetic moment will be the result of circulation of the charge along the rim. Instead of an electric field, the neutron has a negative wave (Fig. 3-b). The structures and the properties of the nucleons will be discussed in greater detail in Chapter 2. Now, it is more important to prove that it is possible to construct nuclei by arrangement of nucleons with disk-like shapes.

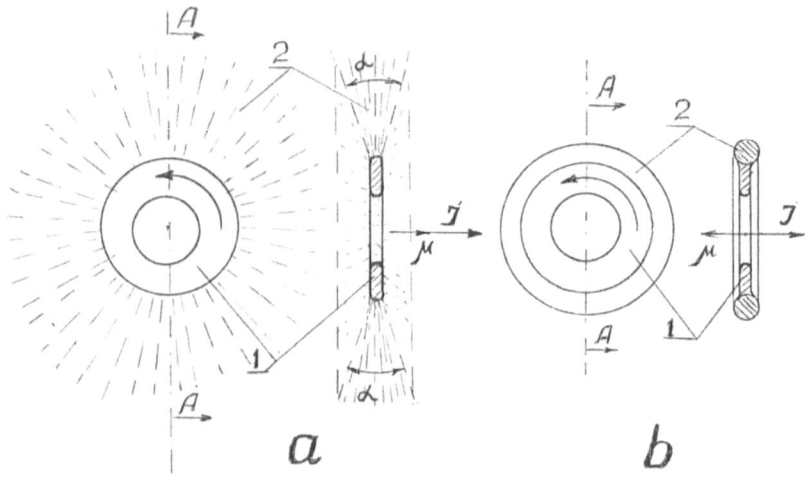

Fig. 3. Scheme of disc-like nucleons.
a) **Proton** 1. Wave; 2. Electrostatic field.
b) **Neutron** 1. Proton wave; 2. Negative wave

1.4. NATURAL NUCLEAR STRUCTURE

The difficult problem of resolving the nuclear structure becomes a simple problem of a disk-like nucleon arrangement in a way that complies with the laws of Nature. The properties of such nucleus must follow from the structure. It is not difficult to guess that the coaxial arrangement of nucleons (proposed in [14]) is a simple solution to the old "insoluble" problem:

a) Via a coaxial arrangement of nucleons Nature has avoided the problem of the "many strongly-interacting bodies". The problem is reduced to the level of "two strongly interacting bodies". The problem is solvable, but the solution will not be an easy one because of uncertainty of the two nucleon potentials, the influences of the neighboring nucleons, and the influence of the "magic" numbers of nucleons. To the credit of many scientists, it must be noted that the role of two nucleon interaction had been anticipated, although not well-understood.

b) The coaxial arrangement of nucleons avoids the direct proton repulsion in the zone of maximum field intensity. The proton repulsion occurs from the weak sides of the fields, where the neutrons can successfully play a role of shields. The supplementary role of the second neutron is to decrease the repulsion by increasing the distances between protons.

c) Only the coaxial arrangement of nucleons provides the possibility for free propagation of the proton electric field.

d) The coaxial arrangement of nucleons is the only possible way of arrangement, which allows for the necessary spin and magnetic moments compensation. Summing up the projections along a "chosen direction" is a type of a scientific conjugation.

Fig. 4 shows a schematic illustration of deuteron (H_1^2). Both disks, the proton (1) with the electric field (5) and the neutron (3) with a negative wave (4) are situated coaxially. α is the angle between the conical surfaces of the proton's electric field. The zone of the "strong" interaction (2) is between the side surfaces of the disks. The electrostatic field of the proton propagates freely with no shielding effects from the side of the neutron. The coaxial direction of the spins allows their summing. Because of the coaxiality and the relative weakness of the interaction between nucleons (only 2,22 MeV binding energy) the magnetic moment of the nuclide is practically equal to the difference of the magnetic moments of the nucleons.

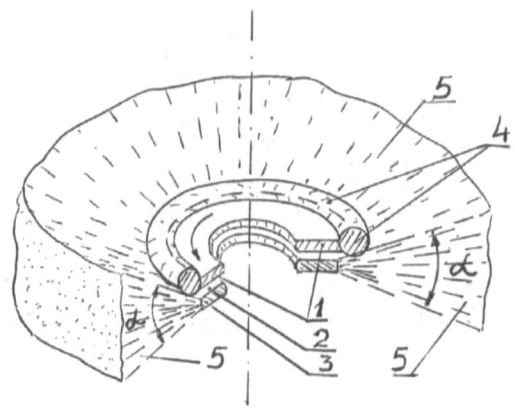

Fig. 4. Scheme of deuteron.
1. Neutron; 2. Zone of strong interaction; 3. Proton;
4. Negative wave of neutron; 5. Electrostatic field of proton

The structure of triton H_1^3 looks like the structure of the deuteron H_1^2 with an additional neutron on the other side of the proton Both neutrons have opposite spin directions.

The structure of He_2^3 (Fig. 5) is reverse to the structure of H_1^3 - one neutron (2) is situated between two protons (3 and 4). Both protons have opposite spin directions. The zone of strong interaction (7) is between the disks. The close proximity of the proton fields (1) predetermines a large zone of proton repulsion (6), which is the cause for the relatively small binding energy (7,72 MeV).

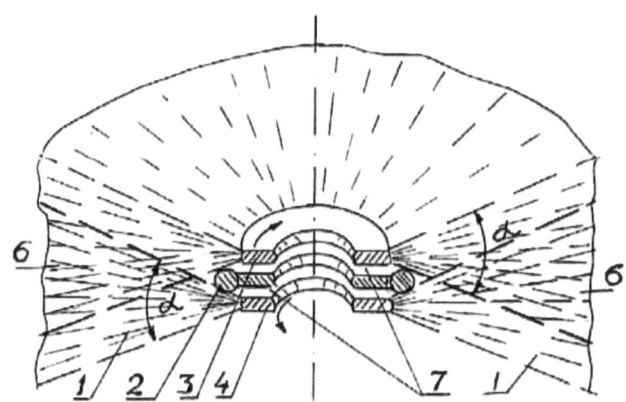

Fig. 5. Scheme of He_2^3
1. Electrostatic field of protons ; 2. Negative wave of a neutron;
3, 4. Proton; 6. Zone of electrostatic repulsion; 7. Zone of strong interaction

Fig. 6 shows a schematic illustration of He_2^4. Both neutrons (2 and 3) with opposite spins are situated between two protons (4 and 5), which also have opposite spins. Two neutrons between the two protons represent a suitable shielding and, as a result, the zone of repulsion (6) is relatively small. Because of opposite spins and small zone of repulsion, the binding energy is impressive (28,3 MeV). The opposite spins of coaxial nucleon arrangement predetermine full compensation of nucleon spin and magnetic moment. As a result, the nuclear spin and nuclear magnetic moment are equal to zero.

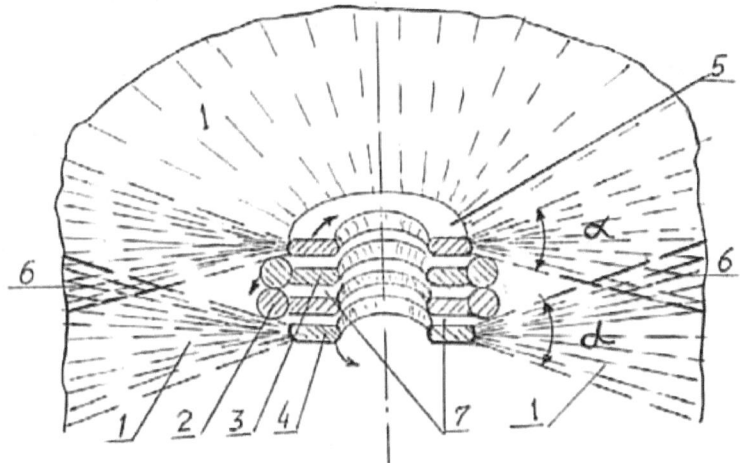

Fig. 6. Scheme of He_2^4

1. Electrostatic field of protons; 2. Negative wave of a neutron; 3. Neutron

Later on, it will be shown that the structure of all nuclei is formed by adding nucleons (or a proton-neutron pair) to these basic nuclides, while observing several rules. These structures are in full agreement with the laws of Physics (and Nature). At first sight, the length of nuclides with middle and heavy masses seems to be unacceptably large. However, for thickness of the nucleons of approximately 10^{-14} cm, the length of the heaviest nuclides will not exceed $2,5 \times 10^{-12}$ cm. This value is comparable to the values of the nuclear diameters, calculated from the well-known equation $R = 1,3 \times 10^{-13} A^{1/3}$ cm. The question of the thickness will receive suitable argumentation later on, after the structure and nature of the nucleons is described.

Chapters 3 and 4 describe how all properties of the nucleus follow directly from the proposed nuclear structure. Clearly, the coaxial arrangement of disk-like nucleons is the only possible structure for the nucleus. *And thus, the first and primary question about the nuclear structure has received its natural solution.* A complete solution needs more details, and above all needs an exposé *of the nature of the disk-like nucleons.* In the next chapter, it will be demonstrated that the concept of the nucleons as closed electromagnetic waves is also a natural solution, supported by countless well-known facts about the wave-matter interaction. This solution is not entirely against the idea of quark structure of the nucleons and even helps to find common sense in the basic concept of quarks.

CHAPTER 2

WAVES AND MATTER

2.1. THE PROBLEM OF WAVES EXISTANCE

In Chapter 1 it was shown how to obtain the natural structure of the nucleus by coaxial arrangement of disk-like nucleons. Such structure avoids the problem of many strong interacting bodies, but the nature of the strong force cannot be explained without revealing the nature of the nucleons. It is not advisable to take in blind faith the idea of quark structure of the elementary particles. Perhaps there are some useful aspects to it, but in general it confuses the problem of the structure of matter by introducing too many particles, too many fields, too many fabricated properties, and too many postulates. The structure of matter should be much simpler. In this aspect, the idea of wave origin of matter looks much more promising. But first we must explore some misleading notions about the nature and even about the existence of electromagnetic waves.

As it follows from the discussions in Chapter 1, a lot of the difficult problems in Physics are created by the scientists themselves. For instance, the notion "wave" is largely used in Physics – from Optics to Wave Mechanics. Nevertheless, a considerable group of physicists consent to the probability interpretation of the de Broglie waves [10], while another group (mainly specialists in the Quantum Electrodynamics field) categorically denies even the existence of electromagnetic waves [11]. To accept the existence of de Broglie wavelength and to deny the existence of de Broglie wave is not very wisely. But what is to be done? Another century of polemics will be too much. The only way to "choose" the right side would be to present more examples, in which the wave nature cannot be rejected. Some well-known problems, such as the reflection from two surfaces, or the diffraction of single electrons, will not be referred to. A problem, which has escaped the attention until now, will be analyzed instead.

2.1.1. Capacitor charging

The capacitor charging is believed to be an ordinary process of energy conversion - first, the potential energy of electrons in a battery transforms into kinetic energy of electrons moving through the wires; then, the kinetic energy transforms into potential energy when the electron stops on the surface of the capacitor dielectric. The process stops when the potential energies of the electrons in the battery and in the capacitor are equalized. *There is no loss of energy during capacitor charging* – no heating of the capacitor plates occurs when the electrons stop! Perhaps as a whole this picture is plausible, but with respect to the details it is wrong because the energy condition in the beginning is not equal to the energy condition at the end of the process: At the beginning of the charging, the potential difference between the plates of the capacitor is zero ($U_c = 0$) and the potential energy of electrons in the battery is $E_p = U_b e$. After switching the key on and off fast, a small part of the electrons obtain kinetic energy $E_k = \dfrac{mw^2}{2} = E_p$, where m is the mass, and w is the speed of an electron. Upon reaching the surface of the capacitor the electrons stop and lose their kinetic energy. Let us assume that the quantity of electrons (q) is small enough in comparison to the capacitance (C), so that the potential difference in the capacitor remains negligible.

$$U_c = \frac{q^2}{C} \approx 0$$

Hence, after stopping on the surface of the capacitor, the potential energy of the electrons becomes nearly zero. *Where did the kinetic energy of the electrons go?* The electrons have no potential energy and there is no heating of the capacitor during charging. Is this a fundamental problem with no answer in the context of Modern Physic? Or is it a violation of the energy conservation law as a new manifestation of the "absurdity" of Nature?

2.1.2 De Broglie waves reality

The twentieth century was abundant in significant discoveries, some of which have had a considerable influence on technical progress. One of the biggest discoveries in science, *the waves of de Broglie,* still remains practically unknown and unrecognized as such.

In spite of the countless proofs, in spite of Shrodinger's wave mechanics, Modern Physics has passed on the death sentence: *De Broglie waves do not exist! Only waves of probability exist!* This is one of the biggest nonsense in science.

What happens during capacitor charging? Under the influence of the potential difference between battery and capacitor, the electrons receive acceleration. In accelerated motion the electrons obtain de Broglie waves – as a matter of fact, each electron "rides" on a wave! The difference between moving and standing still material particles is in the presence and absence of de Broglie waves. For instance, the kinetic energy of electrons is actually the energy of their waves; the mass increase of electrons in motion (known as relativistic effect) is actually the mass of de Broglie waves (this will be proven further in this Chapter); the winding magnetic field at electron motion represents the magnetic field of the de Broglie waves. Usually, when the electron stops, de Broglie wave separates from the electron as an electromagnetic wave. The de Broglie wave is also an electromagnetic wave with some asymmetry of the electrical component. But when the electron stops on the capacitor surface, the de Broglie wave continues on its way through the dielectric to the other surface of the capacitor, where it takes another electron and carries it away to the battery. Therefore the energy conservation law is not violated and through the dielectric between the plates of the capacitor a current of de Broglie waves runs, known as the "displacement current". This is a type of a neutral current carrying energy (mass) and magnetic field. Of course, this magnetic field is just equal to the magnetic field in the conductors. The electron itself is unable to accept or to emit the energy.

The de Broglie waves are one of the fundamental elements of Nature. Every motion is a wave motion. The emission of light is a process of separation of de Broglie waves from oscillating particles. The inertia is a process of formation and destruction of de Broglie waves. As Debay [12] has shown, the heat capacity is directly dependent on the wave propagation.

Although the separation of electrons from de Broglie waves is a convenient phenomenon (electric arc, radio waves emitting, etc.) neglecting the physical meaning of de Broglie waves leads to puzzles. For instance, what will happen during electrical discharge between two electrodes, one of which (the anode) is a superconductor? Being a superconductor, the anode will accept the electrons and will reject their de Broglie waves because of the magnetic field. The rejected waves form a flux of impulses and energy, which can be transmitted at some distance. This is known as the *effect of Podcletnof* [13], and is a puzzle, which interests NASA as a possible antigravity source.

How can the electron (a matter particle) be linked to an electromagnetic wave? This is not odd if we keep in mind that in the pair-production the matter particles are formed from electromagnetic wave and in the particle annihilation the final products are also electromagnetic waves. It is logical to assume, that both processes (pair- production and annihilation) represent only structural changes, during which the wave nature remains unchanged. As it turns out, matter must be a sort of electromagnetic wave. This conclusion immediately explains the basic properties of the matter particles –mass, electric charge, spin and magnetic moment. So the dilemma "matter or waves" has a solution. *At the level of matter everything is a wave!*

The relationship between matter particles and electromagnetic waves was noticed long ago. William Clifford was the first to suggest (in 1876) that matter is composed of pure waves. It is well known, that Schrodinger suggested the electron to be considered as wave-packet. Even Feyneman, in 1945, before totally denying the wave existence, proposed a model of the electron as a composition of "inward" and "outward" waves. The weak point of most similar ideas is the absence of adequate explanation for the nature of light and preconceptions about the form of the matter particles as discussed above.

2.2. NATURE OF MATTER AND WAVES

Until the beginning of the 20th century the physicists were confident in the existence of Ether as a light propagation medium, which remains in the absolute space. The inexactness of this definition contributed to the Ether being expelled from science. Later, under the pressure of the Relativity Theory, in the place of Ether a "mechanical space" was introduced. Today, the notion "physical vacuum" is in use. The problem is not in the name, but in the physical sense of the medium where the electromagnetic waves propagate.

Historically, the problem of the electromagnetic waves arose when they were wrongly interpreted as mechanical waves. The similarity of both types of waves is only in the way of propagation. In principle, the mechanical wave is a sort of motion, and every motion in Nature is wave-like. So the analogy in propagation is quite natural. But the oscillation of the material medium in mechanical waves occurs due to the alternate conversion of kinetic and potential energy of the material particles. The oscillation is gradually dampening because of energy dissipation (i.e., release of de Broglie waves connected with particles). The oscillation in an electromagnetic wave is practically endless – the relic radiation is as old as the Universe. Hence, the electromagnetic waves are not equivalent to the mechanic waves, and *a different type of medium is necessary* for their formation and propagation. A model of this medium was developed and demonstrated in my previous book [14]. The model is in agreement with the laws of Electrodynamics and helps understand the physical sense of all fundamental notions, such as energy, impulse, mass, electric charge, electric and magnetic fields, etc. According to this model, *in the foundation of Nature lies a massless particle with two opposite fundamental charges, or a* **bipole**. The particle is introduced as the origin of substance and charge – as the origin of energy. The fundamental charge (*f-charge,* or simply *charge*) is not equivalent to the electric charge (*or q-charge*). Later it will be shown that a circulation of an f-charge is equivalent to a q-charge. The charge *acts* by means of oscillation and transmits its pulsation directly to the medium. The medium consists of *couples of bipoles, or* **tetrapoles**. The tetrapoles are equivalent to **gravitons** when they are in accelerated motion towards the cosmic bodies. The oscillation of the four charges in the gravitons is mutually compensated, but the repulsion between opposite charges leads to a trend of graviton expansion, which is the cause for *expansion of the Universe with acceleration*. In a bipole, separated from a graviton, the charges oscillate alternatively with a frequency proportional to their energy. This oscillation causes polarization of the medium – a disintegration of gravitons in separate polarized bipoles, which form long *polarized chains*. This is a type of medium polarization, which spreads as an **electromagnetic wave**. At the end of the act of each f-charge oscillation, a chain depolarization and formation of gravitons occur.

2.2.1. Structure of electromagnetic waves

The light consists of two basic structural elements: one oscillating bipole – the equivalent of **photon**, and polarized chains – equivalent to **wave**. This is the reconciliation of the two opposite theories of light. Each chain consists of two *lines* of oppositely polarized bipoles. A polarized bipole in a chain could be considered as a **quasi photon.** The bonds between the quasi-photons define the strength of the chains and are the basic element of electrostatic interaction. The existence of attraction force can only be explained by the strength of the polarized chains! A schematic illustration of a chain is shown on Fig. 7. Each electromagnetic wave consists of many chains united in sheaves. Fig. 8 shows a schematic illustration of a sheaf. The sum of charges in a sheaf forms the *f -charge of a wave*. At each pulsation of each of both photon charges, the oppositely polarized lines of quasi photon move in opposite directions: the linear motion along the sheaves is *the electric motion* and a circular motion around the center of the sheaf is the *magnetic motion*.

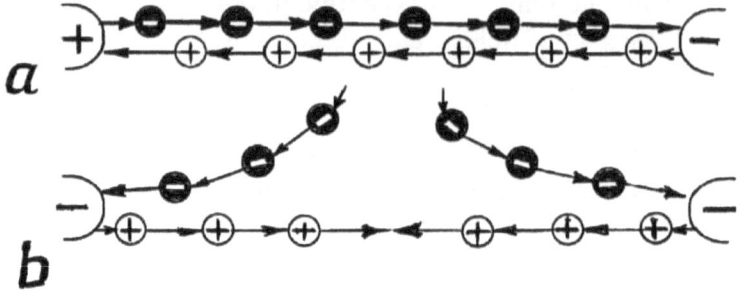

Fig. 7. Scheme of polarized chains
a) Polarized chain between opposite charges (attractive force);
b) Polarized chain between alike charges (repulsive force)

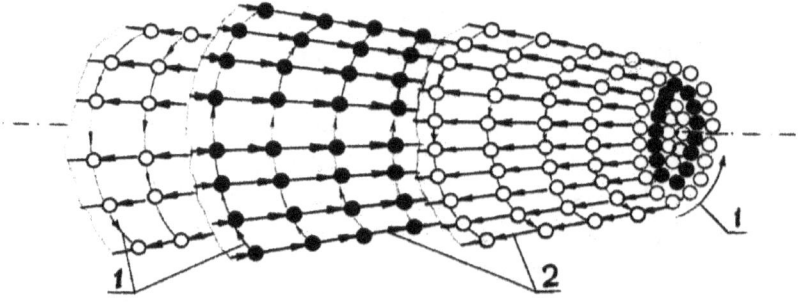

Fig. 8. Scheme of a sheaf of polarized chains
1. Direction of magnetic motion; 2. Polarized chains

The oscillation of a photon charge is the change of its intensity (φ_p):

$$\varphi_p = A_p \exp[2\pi i(vt-\frac{x}{\lambda})]$$

Where: A_p is the amplitude; v is the frequency; λ. is the wave length.

The intensity (φ_w) of the wave charge oscillates by the same frequency:

$$\varphi_w = A_w \exp[-2\,\text{Л}i(vt-\frac{x}{\lambda})]$$

The **force** (F) arises from the direct interaction between the charges and is equal to the product of their intensities:

$$F = \varphi_p \varphi_w = A_p A_w$$

The **action** (D) of the force along the wave length for a time (T) of one pulsation is:

$$D = \int_0^T dt \int_0^\lambda F dx = h$$

Where h is the Planck constant.

From this equation it follows that: $F = h\dfrac{v}{\lambda}$.

The **energy** (e) is the action of the force along the wave length:

$$e = \int_0^\lambda F dx = \int_0^\lambda h\frac{v}{\lambda}dx = hv$$

The **impulse** (p) is the action of the force during the time of a pulsation:

$$p = \int_0^T F dt = \int h\frac{v}{\lambda}dt = \frac{h}{\lambda}$$

Depending on the ratio between the two photon charges, the following forms of waves exist:

a) Symmetric waves

A photon having two equal, opposite charges produces the well-known flat symmetric electromagnetic wave (very often named "light"). The electric component (E) of the microfield stress is proportional to the density of polarized chains:

$$E = -\frac{1}{2\pi i}\cdot\frac{\partial \varphi}{\partial x} = \sqrt{h\frac{v}{\lambda}\cdot\frac{1}{\lambda}}.\exp[2\pi i(\text{и} - \frac{x}{\lambda})]$$

The magnetic component (H) of microfield stress is proportional to the speed of charge-intensity change:

$$H = \frac{1}{2\pi i} \cdot \frac{\partial \varphi}{\partial t} = \sqrt{h\frac{v}{\lambda}} . v . \exp[2\pi i(vt - \frac{x}{\lambda})]$$

As it is shown in [14] the expressions for H and E are in agreement with Maxwell's equations. The ratio H/ E is equal to the speed of light (c): $\frac{H}{E} = v\lambda = c$.

Under the influence of the force (F) the photon moves with acceleration:

$$a = \frac{dc}{dt} = cv$$

The **mass** can be defined as resistance of the medium to the accelerated motion (at de Broglie wave formation):

$$m = \frac{F}{a} = \frac{hv}{\lambda cv} = \frac{h}{c\lambda}$$

Hence, the action (h) expressed through the mass is: $h = mc\lambda$. Since the speed of light is constant, the product $m\lambda = const$. Hence the mass can be interpreted as density of polarized chains along the wave length.

The energy and impulse expressed by mass are:

$$e = hv = mc^2; \quad p = \frac{h}{\lambda} = mc$$

The alternating pulsation of opposite f-charges (of the photon) leads to a fast alternating change of the sign of polarization of the medium. This is why the symmetric wave only has an *instantaneous mass*.

Fig. 9-a shows the well-known schematic of electric (E) and magnetic (H) vectors of a symmetric wave. Fig. 9-b shows a schematic of sheaves arrangement in the same wave.

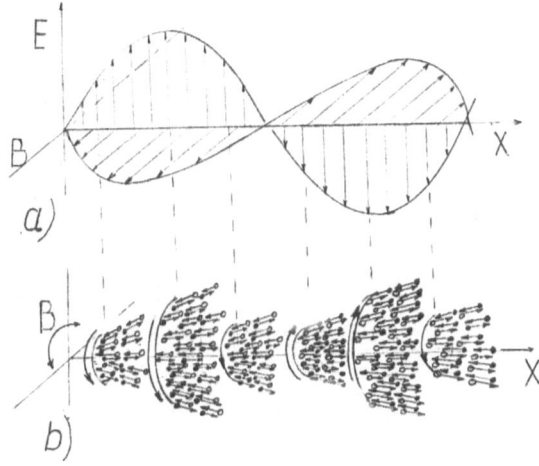

Fig. 9. Symmetric wave
a) Scheme of electric (E) and magnetic (H) vectors;
b) Scheme of sheave disposition

b) Asymmetric waves (Matter waves)

At a collision of a symmetric wave with an obstacle (a mater particle) the photon can be broken in two unequal (asymmetric) bipoles. Because of the asymmetry such bibole can form an asymmetric wave, which has unequal electric vectors. The asymmetry of charges predetermines the wave propagation in a circle. The radius of the circle depends on the degree of the asymmetry of charges. In general, the asymmetric waves are very unstable and shortly they decay in a symmetric wave and asymmetric bipole- a **neutrino**. The asymmetric waves can only be stable as standing waves. Of special importance is the case, when the length of the wave becomes twice as long as the length of the circle, which occurs when there is a big asymmetry. The wave winds twice around the circle. *Such a double-cycled flat wave is actually a mater particle.* The stability of matter particles depends on how exactly the wave length equals two circle (perimeter) lengths, ($\lambda = 4\pi r$), (in order to avoid the interference).

Fig. 10-a shows the vector (E) of electric field stress of an asymmetric electromagnetic wave.

Fig.10-b shows the density (ρ_E) of polarized chains, which is proportionate to the square of the electric vector (E^2).

Fig. 10-c shows the arrangement of polarized chains.

Fig.10-d shows how the basis of this wave is winded up twice to form an electric vector uniform with the circumference. As it was said above, the electric field consists in radial movement of pairs of quasi-photons in polarized chains and the magnetic field consists of transversal (winding) motion of these chains. The stress of the fields is proportional to the density of the chains. The *doubly-rounded wave* can be named **matter wave**. As the matter wave is produced by one charge, the polarization does not change at charge oscillation. Therefore, the matter waves have constant **mass**. Later in this chapter, the problem of matter waves will be discussed in more detail.

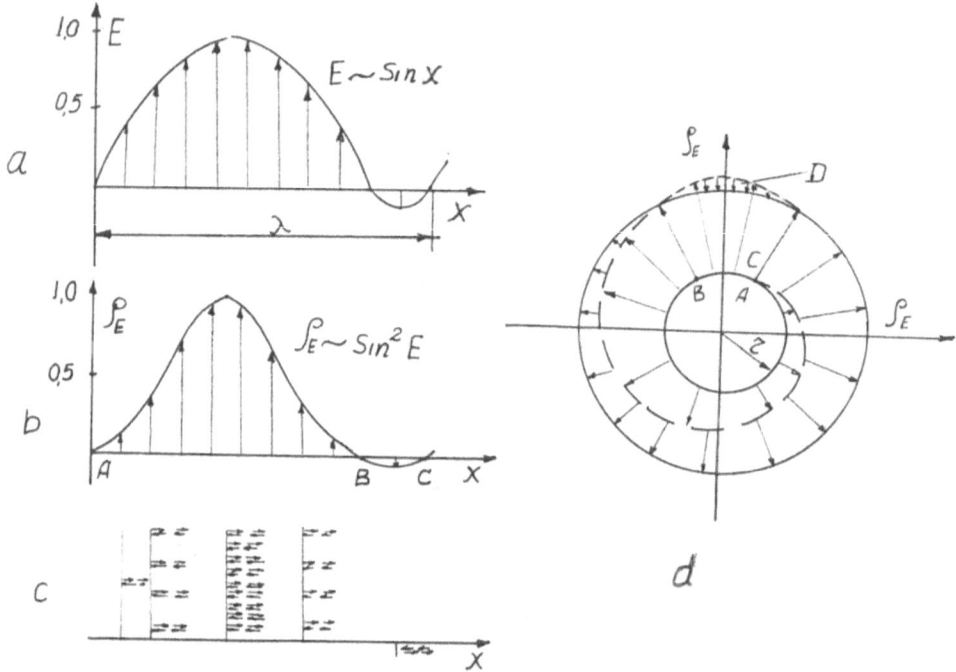

Fig. 10. Matter wave formation
a) Asymmetric wave; b) Density of polarized chains; c) Disposition of polarized chains; d) Matter wave; D Electric vector compensation

Electric charge (q) of a matter particle is the action of the f-charge (φ_m) at its circulation along the wave length:

$$q = k_1 \oint \varphi_m d\lambda = k_1 \oint k_2 \sqrt{h\frac{v}{\lambda}} . d\lambda = k\sqrt{\frac{h v \lambda}{\lambda}} = k\sqrt{hc}$$

The generalized coefficient $k = \sqrt{\dfrac{\alpha}{2\pi}}$ and α is the fine structure constant.

Elementary magnetic flux (ϕ) is the action of matter wave charge (φ_w) during the time of one pulsation:

$$\phi = k' \int_0^T \varphi_w dt = k' \int_0^T k'' \sqrt{h\frac{1}{\lambda t}} . dt = \frac{1}{k}\sqrt{\frac{ht}{\lambda}} = \sqrt{\frac{2\pi}{\alpha}}\sqrt{\frac{h}{c}} = const. = 1{,}37\text{x}10^{-17}.g^{0,5} cm^{0,5}$$

A *"quantum of magnetic flux"* (Φ) exists in Physics as a fundamental constant:

$$\Phi = \frac{h}{2q} = 2{,}082\text{x}10^{-15} Wb = 6{,}82\text{x}10^{-18}.g^{0,5}.cm^{0,5}$$

This value is obtained as magnetic flux of a Cooper pair (note the number 2 in the denominator), and hence is twice smaller then the value of the elementary magnetic flux (ϕ).

$$\phi = \sqrt{\frac{2\pi h}{\alpha c}} = \sqrt{\frac{2\pi h \eta c}{q^2 c}} = \frac{h}{q} = 2\Phi$$

The magnetic monopole does not exist! The magnetic equivalent of the electric charge is the elementary magnetic flux. This is why the product of electric charge and elementary magnetic flux (as a product of charge action) is equal to the action (h) of the force (F).

$$q\phi = k\sqrt{hc}.\frac{1}{k}.\sqrt{\frac{h}{c}} = h$$

The existence of two opposite charges predetermines the possibility of *matter and antimatter* existence. The opposite charges produce matter waves with opposite polarization. The matter and antimatter particles have opposite electric charges. Hence, the electron consists of antimatter relative to the proton. Annihilation is a process of opposite polarized wave interference. Annihilation between proton and electron is impossible because of the difference in their wave frequencies.

c) Compound waves (de Broglie waves)

De Broglie wave is generated when a matter particle moves with acceleration through the medium. The electric component of the de Broglie waves is a superposition of the electric component of a symmetric wave and the constant electric field of the electric charge of the particle. The ratio between the two electric constituents determines the asymmetry of the wave. The properties of de Broglie waves have been discussed above. These properties (mass, impulse and winding magnetic field) are predetermined by the asymmetry of the de Broglie wave. The mass of de Broglie wave is equal to the mass increase of the matter particles in motion. This is easily proven. Fig. 11 shows a matter particle (1) in motion with speed (w) and its de Broglie wave (2).

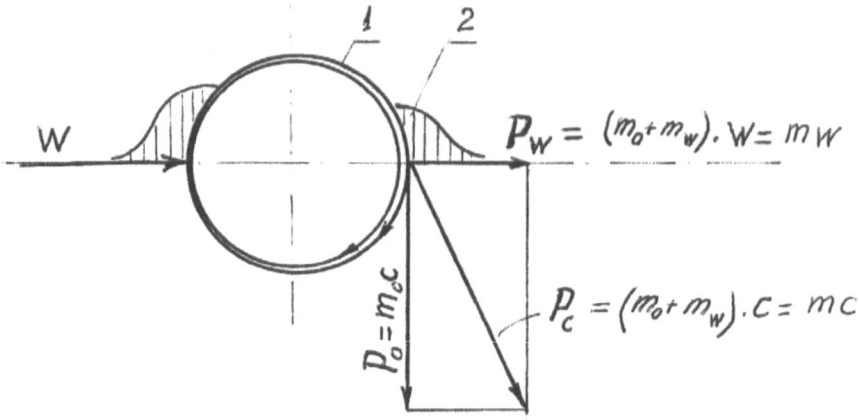

Fig. 11. Mass increase in motion ("relativistic effect")
1. Matter particle; 2. De Broglie wave of particle with mass m_0 in motion at speed w ;

P_w - impulse of particle in motion; P_0 - impulse of matter wave at rest; P_c - geometric sum

The vector of the impulse of mater wave is $p_0 = m_0 c$.

The vector of the particle impulse at motion with speed (w) is $p_w = (m_0 + m_w).w = mw$, where m_w is the mass due to the motion. The geometric sum of these two vectors is $p_c = (m_0 + m_w).c = mc$.

Hence, $(mw)^2 + (m_0 c)^2 = (mc)^2$, or $m = \dfrac{m_0}{\sqrt{1 - \dfrac{w^2}{c^2}}}$;

As $m_w = m - m_o = m_0(\dfrac{1}{\sqrt{1 - \dfrac{w^2}{c^2}}} - 1) \approx m_0 \dfrac{w^2}{2c^2}$, the energy of de Broglie wave is $m_w c^2 \approx \dfrac{m_0 w^2}{2c^2} . c^2 = \dfrac{m_0 w^2}{2}$

Thus, the energy of de Broglie wave is equal to the kinetic energy of the particle.

2.2.2. Absence of wave; Neutrino

The asymmetric photon (bipole with two unequal f-charges), created during asymmetric wave decay, usually has big angular momentum. The centrifugal force of the rotation suppresses the f-charges pulsation and the formation of electromagnetic wave. Such asymmetric photon is known as neutrino. Because of the absence of wave the neutrino has no mass and cannot interact with the matter waves. Neutrino can interact only with the f-charge of the monopole, (which forms the matter wave), but this interaction is strongly restricted by the very small dimension of f-charges, energetic and charges conformity, rotational accordance etc. As a result, the interaction of neutrino with matter particles occurs very rare. The two kinds of the f-charges (proviso named positive and negative) predetermine the existence of neutrino and antineutrino, which rotate in the opposite directions. The energy of the neutrino is confined in its charges. Despite the rotation, the neutrino cannot have a spin and magnetic moment. The big difference in energy is the only distinguishing feature of the different kinds of neutrino (electron, muon, tauon). Transformation of one kind of neutrino into another (neutrino oscillation) is possible in reactions with energy exchanges (elastic and quasi-elastic interactions).

2.3. STRUCTURE OF MATTER PARTICLES

2.3.1. Elementary particles

As it was noted above, the matter wave is a doubly-rounded flat asymmetric wave. The radius of winding is inversely proportional to the asymmetry and to the f-charge intensity. Therefore, the condition for stability of matter particles demands a definite relationship between the f-charges intensity and the asymmetry, which provides just twofold winding of the wave (to avoid the self-interference) and formation of equal electrical vector intensity in circumference. The severe limitations of this condition restrict the stable particles to only two – electron and proton.

2.3.1.1. Structure of the electron

The schematic arrangement of sheaves in a matter wave is shown on Fig. 12. According to Fig. 12, the matter wave is a *wave ring* with tangential arrangement of the polarized sheaves, which form the electric field. The transversal (winding) motion of the polarized chains within the sheaves forms the magnetic moment. This is the structure of a truly elementary particle – it consists of only one wave. Likely, the electron is the only truly elementary particle because all others consist of more than one wave. All properties of the electron ensue from the structure shown on Fig. 12.

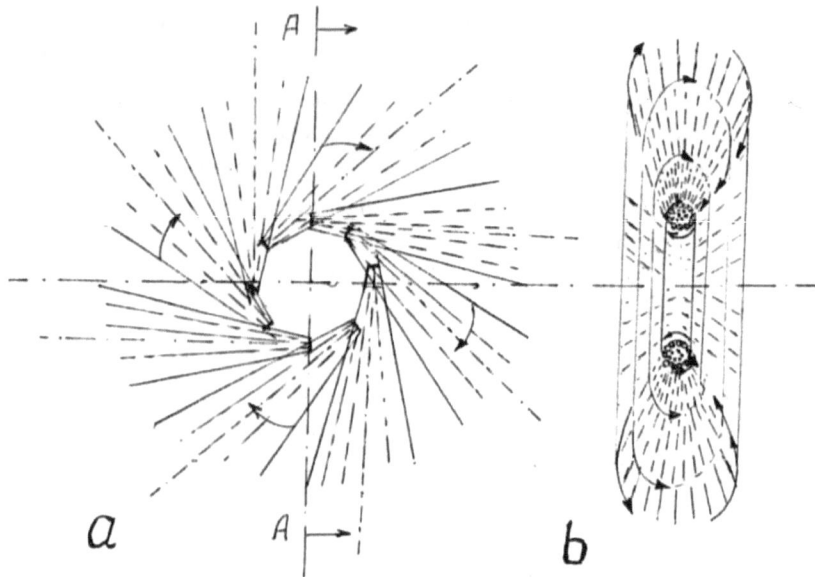

Fig. 12. Structure of electron
a) Disposition of sheaves: b) Cross-section through A-A

I. Electric field structure

Why do not the electron s fall onto the nucleus? Looking at the Physics' history, this was one of the several famous questions, which the Classic Physics was not able to answer, and thus passed on to the Modem Physics. The funny thing is that Modem Physics was also unable to answer that same question. Why does the electron circle around the nucleus? A central force (such as the electrostatic force) cannot produce an orbital motion! Under the central force the motion is predetermined - a straight line towards the center, i.e. towards the nucleus. Then, why indeed doesn't the electron fall onto the nucleus?

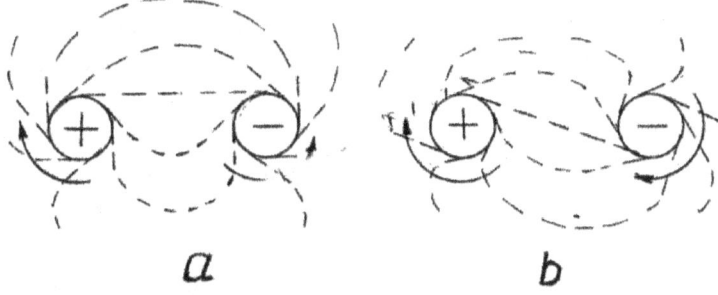

Fig. 13. Scheme of positronium
a) Meta-positronium; b) Ortho-positronium

2. Angular Moment (Spill)

The answer to this question follows from Fig. 12. *The electric force is not a central one!* It looks like central only from macro-distances. Near the elementary particles the *electric force is tangential.* Moreover, the electric field is not a static one. *The electric field rotates together with the matter wave.* Hence the electron is forced to orbit.

Now, what is the difference between the ortho- and meta- states? The Modern Physics explains the difference by the orientation of the spin – a typical trivial explanation of unknown phenomena via unknown properties. On Fig. 13, the schematics of ortho- and meta-positronium (a system of electron and positron) are shown. It is evident from Fig. 13-b, that in ortho-positronium the polarized chains must be slightly longer because the rotation of the electric fields acts against their shortening. Fig. 13-a shows that in meta-positronium the rotation helps for shortening of the chains. As a result, ortho-pozitronium lives longer [2, 3] and the level of its binding energy is by $8,41 \times 10^{-4}$ eV higher than the energy of meta-positronium

The usual definition of the spin is: "Intrinsic angular momentum"; sometimes with the supplementary explanation. "The spin has quantum origin. The angular momentum is not due to transference of the particle as a whole [3]". Scarce information indeed!

In Mechanics, the angular momentum (J) is related to a motion with speed (w) of a body with mass (m), along a circumference with radius (r), and is equal to the product of the impulse ($p = mw$) and the radius, or: $J = pr = mwr$. The electromagnetic wave (or the photon) has an impulse ($p = m_i c$), but it does not have an angular momentum. Hence the photon cannot have a spin. The Modern Physics attributes to the photon a spin equal to η. This is a fictitious spin and follows from the possibility of photon impulse transformation in angular momentum when the photon is divided in two pieces and each piece circulates, producing a spin of $\frac{1}{2}$ η. This conclusion is a direct proof of the wave origin of matter particles.

The essence of the spin follows from the wave properties of matter. The impulse of an electromagnetic wave is a result of pulsation of photon charges. The impulse produced at pulsation of both charges and transmitted to the wave is $p = m_i c = \dfrac{h}{\lambda}$, (where m_i is an instantaneous mass). Hence, the impulse produced by one pulsation of one charge is equal to $\dfrac{h}{2\lambda}$ and it does not depend on the charge intensity (φ). But the matter wave is double and hence its impulse (p_w) is twice as much, or: $p_w = mc = 2\dfrac{h}{2\lambda}$. Hence: $mc\lambda = h$. At matter waves $\lambda = 4\pi r$ and $4\pi mcr = h$.

The angular momentum, or spin of the wave, must be $J = mcr = \dfrac{h}{4\pi} = \dfrac{\eta}{2}$.

Hence the spin of the electron is equal to $\frac{1}{2}$ η.

The angular momentum (spin) of electron in orbit is equal to the product of the de Broglie wave impulse and the radius of the orbit. The wave impulse is produced by pulsation of one photon ($p_w = \dfrac{h}{\lambda}$). The length of orbit (l) is equal to the product of the principal quantum number (n) and wave length: $l = n\lambda = 2\pi r$, or $\lambda = \dfrac{2\pi r}{n}$. Then, the impulse can be expressed as: $p = mw = \dfrac{h}{\lambda} = n\dfrac{h}{2\pi r} = n\dfrac{\eta}{r}$

Hence, the spin of the electron orbit is: $J = mwr = n\eta$.

3. *Magnetic Moment*

According to the Classical Physics [2, 3], a circular motion of electric charge (q) with speed (w) forms a magnetic moment $\mu = \dfrac{IS}{c}$, where $I = \dfrac{qw}{2\pi r}$ is the current, and $S = \pi r^2$ is the surface of the circle.

Hence:

$$\mu = \frac{qw\pi r^2}{2\pi rc} = \frac{qwr}{2c} = \frac{qmwr}{2mc}$$

Where m is the mass of the particle.

The product $mwr = J$ is the spin. Since the spin of the electron is equal to $\frac{1}{2}\,\eta$;

$$\mu = \frac{q\eta}{4mc} = \frac{1}{2}\,\mu_B, \text{ where } \mu_B \text{ is the Bohr magneton.}$$

However, it is a well-known fact that the magnetic moment of the electron is equal to the Bohr magneton! The problem is not in the Classical Physics' model, and quantization is not the way for its solution. The answer is very simple. As it was said above, the winding magnetic field is produced not by the charge (q), but by the wave, and the electron is made up of a double wave. Hence, as it follows from the equation above, the magnetic moment of the electron should be twice as much as a single wave. So the equality of the magnetic moment of the electron to the Bohr magneton is another direct proof of the wave structure of elementary particles. The negligible difference (approximately 1,16%) of the electron's magnetic moment from the Bohr magneton is due to the difference between the radii of charge and mass circulation.

2.3.2. Compounded matter particles

Due to of some unknown reason, Nature has only created the electron as one-wave (elementary) particle. All of the other matter particles are compound, i.e. their structure consists of several waves. Evidently, Nature uses the same principle of spin and magnetic moment compensation both, in compound particles, and in the nuclear structure. Hence the coaxial arrangement of flat doubly-rounded waves is the most probable structure of the elementary particles. The binding force between the waves has the same origin as in the nucleus – wave interference (strong interaction).

Probably, in the world of particles, wave with energy 35 MeV plays a role of a basic structural element (wave quantum). It is easy to show that the masses of the long-living particles are divisible by 35 MeV. For instance:

Muon (mass 105,7 MeV) - 3 x 35 = 105;

Pion (mass 139,6 MeV) - 4 x 35 = 140;

Kaon (mass 493,7 MeV) - 14 x 35 = 490.

The differences between the masses of Hyperions and the mass of the proton (938 MeV) are also divisible by 35 MeV:

Λ (mass 1115 MeV) - 938 MeV = 177 $\approx$ 5 x 35 =175;

Σ^+ (mass 1189 MeV) – 938 MeV = 251 $\approx$ 7 x 35 = 245;

Ξ^+ (mass 1321 MeV) – 938 MeV = 384 $\approx$ 11 x 35= 385.

The differences between the actual and calculated masses are below 1% - precission unattainable in other models of elementary particles. A difference within 1% is quite normal in view of the mass defect in strong interaction. The properties of the particles are also in agreement with the number of 35 MeV waves. The particles with even number of waves (the mesons) have zero spin and do not have magnetic moment because of their compensation at opposite direction of circulation. When the number of opposite circulating waves is not equal, the spin of the particle is equal to the whole number η. The particles with odd number of 35 MeV waves have magnetic moment and fractional spin. The analysis shows, that the structure of all particles can be presented as groups (bunches) of 35 MeV waves, but some positive elements of the quark structure suggest the idea of the existence of several types of waves.

2.3.2.1. Qark Analogy

From a common sense point of view, the idea of quark structure is not only extravagant (fractional charge, color, charm, etc.), but also quite improbable. First of all, the indefiniteness of the quark's mass is totally unacceptable because the mass is the main and the most measurable characteristic of all matter particles. In general, the mass is an add-on characteristic. The mass defect in strong interaction cannot exceed 1% and in hypothetical cases it cannot exceed several percents. Thus the strong interaction between quarks cannot be a justification for them being mass-indefinite. Secondly, the quarks are not elementary particles. In the scale of Nature, one simple substance must lie in the foundation (beneath the two basic particles- electron and proton). A dozen of particles and fields are not a suitable basis for Nature. The question about the origin of all these particles and fields cannot be neglected. Dozens of different substances?..._Where are we going?

Hence the existence of quarks as matter particles is impossible. The only reasonable possibility is: *The quarks must be interpreted as bunches of waves.*

The scientists search for quarks in colluders. But what happens in accelerators? → Formation of de Broglie waves. What happens in colliders? → De Broglie waves crashing and formation of matter waves. Depending on the energy and asymmetry of crashing the apparatuses will register stable particles, unstable particles, resonance and quarks. *The particles do not exist in advance as quasiparticles in imaginary particle field.* They are product of electromagnetic wave crashing. *The bred is not made from bits and pieces, but it could be broken into pieces.* Hence the proofs of the quarks existence are actually the proofs for their wave nature.

An analysis in my previous book [14] shows, that the equivalent of **d**–quark consists of one **d**-wave, or of a **bunch** of **d**–waves. Each **d**-wave has electric charge (q), spin ½ η, and mass 35,2 MeV. Each bunch of **d**-waves behaves as one wave with a common spin ½ η, a common charge (q) and a common magnetic moment. The mass is only a sum of the masses of waves.

The **muon** consists of a bunch with 3 **d**–waves. This construction provides a mass of 105,6 MeV, a spin ½ η and a magnetic moment in accordance with the mass and spin values.

The equivalent of **u**–quark is one **u**-wave, or a bunch of **u**-waves. Each **u**–wave is electro-neutral (a sort of standing wave), has a spin of ½ η, a magnetic moment and a mass of 33,8 MeV. The bunch of **d**-waves behaves as one wave with a mass equal to the sum of masses of waves in the bunch.

The charged **pion** consists of a bunch of 3 **d**–waves and one **u**–wave with opposite spins. This structure provides a mass of 139,4 MeV, a spin of zero and absence of magnetic moment. The instability of the neutral **u**–wave causes the decay of the pion. The **u**-wave decays into a neutrino and de Broglie wave. The remaining 3 **d**-wave forms a muon. The neutral **pion** (mass of 135 MeV) consists of 2 oppositely circulating **u**-bunches, each comprising of 2 **u**-waves with opposite polarization. This structure predetermines the mass of 4 x 33,8 = 135,2 MeV, a spin and a magnetic moment of zero, and a shorter half-life time because of the prompt annihilation of the oppositely polarized waves.

The equivalent of **s**–quark is a bunch of electrically charged **s**-waves having mass of 36,7 MeV. The charged **kaon** (mass 493,7

MeV) consists of one **s**-bunch (of 7 **s**-waves) and another, **u**-bunch (of 7 **u**-waves) with opposite spins (calculated mass of 493,5 MeV). The neutral K_s^0 **kaon** consists of one **s**-bunch (of 3 **s**–waves) and one **d**–bunch (of 11 **d**–waves). K_L^o consists of 1 **s**-bunch (7 **s**-waves), one **d**–bunch (3 **d**-waves) and one **u**-bunch (4 **u**–waves). These structures provide a mass of 397,3 MeV (actual mass of 397,4 MeV) and explain the difference in the modes of decays.

The equivalent of **c**-quark is a bunch of waves equal to the waves of 3 kaons. It is not difficult to find bunches equivalent to other quarks. By this set of waves and bunches of waves, it is possible to explain the structure and the properties of all particles. The structure of resonances is more or less undetermined because the half-life of $10^{-23} - 10^{-24}$ sec is practically equal to the time of one pulse of the f-charge and the formation of matter wave is illusionary.

2.3.3. Structure of nucleons

Proton. If the idea of a 3-quarks structure of baryons is correct, proton must consist of 3 bunches of waves. According to the study of the *formfactor,* inside the proton there is a negative charge. Thus one of the bunches must be negatively charged and must circulate in opposite direction. This structure provides an electric charge (q), a spin of ½ η and a magnetic moment of approximately 3 times as much as the nuclear magneton (μ_N). The actual value of the proton's magneton is 2,79 μ_N. The difference is due to the strong interaction between the bunches. The total number of waves is 27 and there are different possible wave combinations for the structure explanation. The difficulty comes from the uncertainty of the mass defect as a binding energy between the bunches. Anyway, the mass of 27 waves (27 x 35,2 = 950,4) is by 12,1 MeV bigger than the mass of the proton (938,3 MeV). The difference of approximately 1% is quite acceptable as binding energy of bunches in the proton. From the outside the schematic structure of the proton must look like the schematic structure of the electron, shown in Fig. 12. Only the radius of the wave circle is considerably smaller. To the total mass of proton corresponds a wave radius of approximately 10^{-14} cm.

Neutron. The neutron is only electrically neutral. If the hypothesis about the charge independence of the strong interaction is correct, the basis of the neutron structure must be a proton. It is possible that the neutron is a proton with a small asymmetric negative wave. This negative wave consists of the wave of an electron and its de Broglie wave, situated in proton potential (electric) field. The bond between the two oppositely charged waves is pure electromagnetic. The weakness of this bond is due to the negative wave trend towards straightening because of its small asymmetry. So the structure of the neutron is much like the preliminary structure (see Fig. 3-b).

It is well known, that the spin of the neutron is equal to ½ η, and this must be the spin of the positive wave. At β^- decay occurs formation of electron with a spin equal to ½ η, and the spin balance of the reaction requires a spin of ½ η to be imputed into neutrino. However, the mass-less neutrino cannot have angular momentum. The neutrino have only an f-charge, which is able to produce impulse and angular momentum at eventual collision with matter particle. Evidently the spin of the neutron might not be exactly equal to ½ η, because the negative charge must form also a small angular momentum (less than 10^{-3} η, as the frequency of negative charge pulsation is approximately 2000 times less than the frequency of proton wave pulsation). But at electron formation (at β^- decay), the pulsation of this negative charge is able to produce an angular momentum equal to ½ η, because the radius of the electron is approximately 2000 bigger than the radius of the proton.

In summary, the structure of the nucleons is presented in limited volume in this Chapter as it is only necessary for understanding the nature of nuclear forces and explaining the nuclear properties. Probably in the future it will turns out, that not all assumption will be absolutely correct (the structure of the medium for instance), but they are enough good as a beginning. The results obtained on their base are in agreement with the laws of electrodynamics and allow to give sensible explanation of the basic physical notion, such as wave, mater, energy, impulse etc.

CHAPTER 3

NUCLEAR PROPERTIES

3.1. PRINCIPLES OF NUCLEON ARRANGEMENT

The properties of the nucleons impose the following rules of arrangement:

- *Proton/neutron sequence* is dictated by the necessity for of the proton setting wide apart, in order to reduce the electric repulsion.

- *Proton/neutron couple formation* is dictated by the necessity of spin compensation – a condition for a stronger bond.

- *Maximum spin and magnetic moments compensation* within the body (**frame**) of the nucleus.

These principles of nucleon arrangement are mandatory and predetermine the following structure. Each nucleus consists of *two branches*, herein referred to as *left* and *right*. Each branch in even/even nuclei contains one half of the nucleons. In even/odd nuclei one of the branches (herein referred to as left one) contains one proton more (*extra proton)* than the other (the right) branch. In odd/even nuclei one branch (herein referred to as right one) contains an *extra neutron.* In odd/odd nuclei the left branch contains an extra proton, and the right branch contains an extra neutron. These *extra nucleons* form the *nuclear asymmetry* and exert very strong influence on the nuclear properties. In each branch the protons and neutrons have different collective behavior. Therefore it is advisable to distinguish two rows in each branch – a *proton row* and a *neutron row.* Thus, there are four rows in each nucleus – two proton and two neutron rows (although all nucleons are arranged in one line). The spin of all nucleons in a row has the same orientation. In each branch, the spin of protons is opposite to the spin of neutrons. Hence the spin compensation between protons and neutrons occurs within the branch. Further, shielding and compensation of nucleon magnetic moments occurs in each branch. In this book, the direction of proton rotation (the spin) in the right branch is provisory accepted to be "from us" - opposite to the proton rotation ("to us") in the left branch. The two branches meet in the *middle* of the nucleus. A final spin and magnetic moment compensation occurs in the middle of nucleus. Usually, the extra nucleons are disposed in the middle of the nucleus.

3.2. NATURE OF NUCLEAR FORCES

3.2.1. Attractive force of strong interaction

The attractive force between the nucleons cannot be brought from outside. It is well known that approximately 1% of the mass of nucleons is lost as binding energy (mass defect). It was discussed in Chapter 2 that the mass of matter waves is proportional to the density of the polarized chains. Hence the mass defect is a result of the density decrease because of the interference between oppositely polarized chains in the matter waves. These chains are situated on the inner side of the proton wave – nearer to the center. Hence the force of attraction exists only between the proton waves. (Note that the main wave in neutrons is a proton wave). The influence of spin orientation on the force can be seen on Fig. 14. The ties between sheaves in matter waves with a parallel spin (shown schematically on Fig. 14-b by only one polarized chain) are larger and weaker. The binding energy is smaller, but the tie is more flexible and steady to stress. When the spins have opposite directions (see Fig. 14-a) the tie is shorter and stronger. In this case the interference goes along with a small mutual turning of the nucleons. The tie between nucleons is strong and rigid in the axial direction; in-plane perpendicular to the axis the tie is more flexible. This flexibility allows oscillation and axial displacement of nucleons.

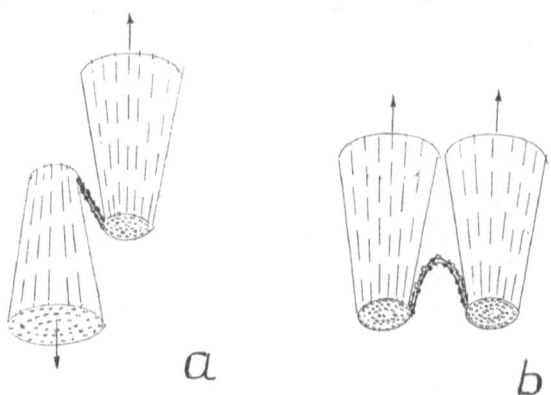

Fig. 14. Scheme of ties between sheaves
a) Opposite directions of nucleon spins; b) Parallel direction of nucleon spins

At a certain condition (low temperature) the tie between waves can overcome the electrostatic repulsion, as for instance, in the formation of an electron Cooper pair. Formation of such pair between two protons is impossible, because the area of wave interaction is smaller than that in electrons (approximately million times) and the force of attraction cannot overcome the electrostatic and magnetic repulsion. But this tie could be realized when a part of a proton electrostatic field is connected by the neutron negative wave, as it occurs in the case of the nuclei of He_2^3 at very low temperatures.

3.2.2. Repulsive forces

- *Proton stress.* The repulsive forces have an electrostatic character. The axial arrangement of protons reduces considerably the repulsion between them, but with the proton number increasing the stress of the net electrostatic field (*or proton stress*) increases. The proton stress is the cause for the electron capture, β^+ -decay, α - decay, proton decay and for the fission of the heaviest nuclei. The role of the neutron is not only to shield the positive electrostatic field, but to space out the protons and so to reduce the local intensity of the field (see Fig. 5 and Fig. 6). The repulsive force of the proton stress exists even in the richest in neutrons nuclei.

- *Neutron stress.* The neutrons play a very important role in nuclei as shields and as internal constraints between protons. Although there are no nuclei without neutrons, there are also no nuclei which consist only of neutrons. In addition, in the stable and long living isotopes, it is impossible to locate more than two neutrons side by side. Evidently, despite the electro-neutrality, there is a strong enough repulsion between the neutron negative waves (*neutron stress*). Two neutrons can be in permanent tie only when they are located between two protons and certain parts of their negative waves are bound with the positive electric fields of neighboring protons. In this state the negative wave is practically bound by two protons and consequently the neutron stress (repulsion) decreases. The neutron stress exists in all nuclei having at least two neutrons side by side. If the stress is strong enough, one of the neutrons must undergo a β^- decay.

3.2.3. Destructive force of nuclear asymmetry

The nucleon in a nucleus participate in different form of collective motion (see below) of wave character. Each wave motion includes different possibilities for interference and brings different consequences for the nuclear stability. It is well known that the interference is very sensitive to any asymmetry. In a nucleus having extra nucleons the uncompensated wave motion of the extra nucleons forms a destructive inertial force. This force is the cause for the limited number of stable odd/odd nuclei (only 4) and smaller number of odd/even and even/odd nuclei in comparison with the number of even/even nuclei.

3.3. MOTIONS WITHIN NUCLEI

The axial nucleon arrangement and oppositely acting forces predetermine different kind of individual and collective motion of the nucleons in nuclei.

3.3.1. Oscillation (the nature of magic numbers)

In the stable nuclei as a whole, the attractive and repulsive forces are in equilibrium, but inside the nuclei the dispersion of energy leads to permanent axial oscillation of the nucleons. Each nucleon being subjected to action of opposite forces must oscillate between its two neighbors and together with its neighbors must take part in a more composite form of oscillation. *Each row of nucleons looks like a multi-section spring pendulum.* The frequency and amplitude of oscillation of each nucleon depend on the characteristics of its neighbor's oscillation. The oscillations in each branch spread as two parallel waves (proton and neutron waves) from the middle of the nucleus to both ends of the branches. It means that there are usually four waves in each nucleus spreading in pairs in opposite directions counterbalancing each other. The lack of balance between the waves is the cause for arising of an inertial destructive force (see above) which reduces the stability of nuclei.

The wave length (λ) depends on the number of nucleons because all nucleons take part in the oscillation. For a certain number of nucleons the wave becomes standing – i.e., the amplitude (A) of nucleons on both ends of the wave becomes zero. The number of nucleons (N) of each type in one branch at which the wave becomes standing can be determined from the equation:

$A = A_{max} \sin(\frac{x}{\lambda}) = 0$. Evidently: A = 0, when $\frac{x}{\lambda} = \frac{\pi}{2} = \frac{k}{\lambda} N$

or $N = \frac{\pi\lambda}{2k}$. Let us accept the idea of standing wavwe lenth quantization and express the wave lenth as $\lambda = (const)^n$, where n can only be an integer. Then an analysis shows that the number of nucleons in each row which can form standing waves must be equal to $N = 1{,}56 \times 2{,}5^n$.

So, for $n = 1$, N = 3,93 ≈ 4, In the nucleus: 4 x 2 = 8;
for $n = 2$, N = 9,80 ≈ 10, In the nucleus: 10 x 2 = 20;
for $n = 3$, N = 24,5 ≈ 25 (or 24), In the nucleus: 25 x 2 = 50(or 48);
and for $n = 4$, N = 61,5 ≈ 62 (or 61), In the nucleus: 62 x 2 =124 (or 122);

(Multiplication by two is because there are two branches and hence two standing waves in a nucleus). Thus the number of protons and neutrons in nuclei with standing waves are 8, 20, 50, 124, and these numbers represent the majority of the *magic numbers*. If this estimation is correct, it would be very likely that the nuclei with numbers 48 and 122 would possess "magic" force. The rest of the magic numbers can be obtained as superposition, for instance 28 is a sum of 8 and 20, and 82 is close to the sum of 50 and 28. This logic leads to a conclusion about the possible existence of more magic numbers of superposition, for instance number 16 (8 + 8), numbers 56 (48 + 8) and 58 (50 + 8), number 78 (50 + 28) etc. The "magic" of these numbers, obtained through superposition, should be manifested by less intensive oscillation. Their "magic force" would not be as strong, but it would influence the nuclear properties. Perhaps number 2 is not a "magic" one, because a wave cannot consist of one nucleon. That is why the nuclei of H_1^3 and He_2^3 having 2 neutrons and 2 protons respectively cannot be related to the group of "stout" nuclides. Moreover, H_1^3 is radioactive. But evidently a single nucleon is not predisposed to a strong oscillation. Thus, the cause for the enormous stability of He_2^4 is mainly a consequence of favorable combination of electric field shielding and spin orientation.

The nucleon oscillation is accompanied by a *charge oscillation*. During the oscillation, the negative waves of a neutron temporarily pass to the neighboring protons. This charge displacement is equivalent to an electric current. The charge oscillation is easy to induce because it needs less energy, and gives an illusion of a nuclear superconductivity.

The *charge exchange* occurs mainly in nuclear reactions, when the passage of negative wave becomes irreversible. Because of the opposite spins of the neighboring proton and neutron, usually the charge exchange imposes a nucleon spin changes.

3.3.2. Rotation

Nucleons rotation – spin. As discussed in Chapter 2, the matter existence in the Universe is due to the f-charges circulation. The spin of the nucleons is a consequence of a type of wave rotation. The possibility of spin compensation at coaxial arrangement of nucleons is a condition for the nucleus existence. The nuclear spin is a result of the nucleon spin balance. Any violation of the spin balance reflects on the energy balance. Energy from outside can change the spin direction of one or several nucleons, which will reflect on the nuclear spin and nuclear magnetic moment. Spin direction changes occur very often in nuclear reactions.

Nuclear axial rotation. At f-charge pulsation, a reactive force arises and causes a nuclear rotation in the direction opposite to the spin direction. All nuclei having spin different from zero must rotate around the nuclear axes. The speed of this rotation depends on the nuclear momentum of inertia. The influence of this speed on the nuclear magnetic moment in light nuclei could become observable.

Nuclear transversal rotation. A part of the spin compensation along the branches occurs in **pn** couples. The final spin compensation of the unpaired neutrons occurs in the middle of the nucleus, but the opposite spin directions of unpaired neutrons in both branches form a force pair with certain momentum able to set the nucleus in slow rotation around an axis perpendicular to the nuclear axis.

Nucleon rotator (nuclear "orbits"). An individual nucleon cannot circulate in orbit around the nucleus like an electron in an atom. First of all, there are no central attractive forces acting outside the nucleus. On the contrary, there is a strong repulsive force outside of the nucleus for the proton in orbit. A neutron could only slowly crawl on the nuclear surface, if the binding force is not so strong and if there is a moving force (for the crawling) outside the nucleus. Secondly, any removal of a nucleon will destroy the nucleus. If a proton is pulled out it could not any more return inside the nucleus. Hence a circulation of a nucleon in orbit at a noticeable distance away from the nucleus is impossible! But the analysis of nuclear properties shows that the behavior of some nucleons is similar to the nucleon in orbit. What is the reason? The most probable explanation is:

Under the influence of the inertial force of oscillation (the destructive force of asymmetry) an axial displacement of the extra nucleons can take place. The nucleon does not leave its place in the nucleus as a whole. Only the center of the nucleon is displaced from the nuclear axis, probably at a distance of one nucleon radius. The ties of the strong interaction (polarized chains interference) are extensible in the plane of the matter wave. The displaced nucleon behaves as a rotator. The nucleon rotation forms a de Broglie wave, which possesses all the properties of an electron orbit in atoms. As a rule, the extra nucleons are in the center of the nucleus, where the bonds are weaker and more stretchable. The center of the nucleus is the place of oscillating energy balance, and a part of the excess energy in rows with extra nucleons goes towards the extra nucleon displacement. Hence the nucleon displacement is in a way a reaction against the asymmetric destructive force. This determines a mutual dependence between the energy of de Broglie wave (spin energy) and energy of oscillation in a nucleus. The deviations of this dependence are the basis of *nuclear isomerism.*

3.4. NUCLEAR PROPERTIES

3.4.1. Nuclear spin

The spin problem in the nucleus has two parts: the means of nucleon spin compensation and the means of nuclear spin formation.

The nucleon spin compensation means that there is the same number of nucleons with opposite spin direction (or opposite angular momentum). As it has been established in this book, only the axial arrangement of nucleons ensures the possibility of a full nucleon spin compensation. Nucleon spin compensation means maximum binding energy and stability of the nucleus. That is why the nucleon spin compensation is a fundamental trend in nucleus formation. Spin compensation takes place in the frame of the proton/neutron (**pn**) pair, in the frame of each of the branches of nucleus, and in the frame of the nucleus as a whole. Full spin compensation occurs only in even/even nuclei. In nuclei with odd numbers of nucleon the spin of the extra nucleons as a rule remains uncompensated.

The nuclear spin formation. Because the spin is a result of the pulses of charges it does not depend on the binding force (strong interaction). There is no difference between the spins of a free nucleon and the spin of a nucleon inside a nucleus. In the ground state, the nuclear spin is a sum of the extra nucleons spin. The spin of the extra nucleon includes also the spin of de Broglie wave ("orbit") whose energy is on the order of MeV. The extent of extra nucleon shifting (the radius of the "orbit") is limited by the possibility of polarized chain extension (force of strong interaction). The extra nucleon rotation sets the nucleus in a slow rotation, which is equivalent to a rotation with increased effective mass. So with the rotation speed increasing, the length of de Broglie wave decreases rapidly and then more wave length can be set on a length of the "orbit". At these conditions each impulse of the charges of de Broglie wave produces supplementary angular momentum of one η, and the spin of "orbit" becomes equal to the number of de Broglie wave lengths in it (as for electron orbits in atoms). Hence the cause for the elevated values of the nuclear spin in the ground level of the nucleus is the vigorous oscillation of the nucleons in asymmetric nuclei, which leads to an extra nucleon displacement. That is why there is a mutual dependence between the energy of "orbit" (spin energy) and the energy of oscillation. The "orbit" of a proton has a magnetic moment which must be taken into consideration in nuclear magnetic moment calculation.

In the exited state of a nucleus, spin changes of several nucleons can occur. A change of the spin direction of one nucleon leads to the nuclear spin increasing with η. The change of the spin direction of a nucleon needs energy. Depending on the nuclear structure, this energy is usually in the range from one to several KeV.

In the excited state of a nucleus, more nucleons (besides the extra nucleons) could be shifted into "orbits". The number of the shifted nucleons and the number of de Broglie wave lengths in each "orbit" (which determines the spin of "orbit") depend on the energy of excitation. Probably, the exited state of a nucleus with many shifted nucleons is the basis of the *Independent particle model* or *Collective model* - if each shifted nucleon (having spin bigger than ½ η) is considered as "valent" one, and the rest of the nucleons are considered as "core".

The general conclusion is: *The nucleus strives to overcome the asymmetry, to reduce the destructive force by sending off the extra nucleon into "orbit".*

3.4.2 Nuclear magnetic moment

The magnetic moment of a nucleus is a result of addition, compensation and shielding of the magnetic moment of nucleons and magnetic moment of "orbits". Each nucleon shields to some degree the magnetic moments of its neighbors. The addition occurs along each of the nuclear branches. A compensation of magnetic moments of both branches occurs in the middle of the nucleus. The repulsive force which arises in the compensation helps the sifting of extra nucleons. As the magnetic moment is a property of matter wave (a manifestation of a tangential wave rotation), there are several peculiarities:

- In the processes of addition and compensation some elements of wave interference are necessarily included. Hence as a rule, the result of addition will be different from the arithmetical sum.

- The magnetic field as a part of the wave takes part in the strong interaction, (which is a form of wave interference). Hence the magnetic moment of the nucleons must depend on the binding energy. *The magnetic moment of a nucleon in the nucleus is always less than the magnetic moment of a free nucleon.*

- The rotation of the nucleus as a whole makes thicker or thinner the polarized chains and thus influences the magnetic moment of the nucleons.

- The coaxiality and the the planes coincide of nucleon and "orbital" rotation allows the summation of magnetic moment of extra nucleon and magnetic moment of its "orbit". Probably the "orbit" of the neutron has no magnetic moment.

For a precise magnetic moment calculation, each of these peculiarities needs a proper quantitative expression. Until then only approximate, qualitative estimation would be possible.

3.4.3. Electric quadruple moment

Strictly speaking, the existence of a quadruple moment in the nuclei is a misconception. In Physics the electric quadruple moment is a static system of two dipoles. Where does the negative charge in a positively charged medium come from? How to explain the existence of quadruple moment in deuterium (H_1^2) which contains only one positive charge? Evidently in the nucleus the role of the negative charges is played by the negative wave of the neutrons. The axial arrangement of protons and neutrons consequently corresponds to a formation of dipole, quadruple, octuple and even the moments of higher orders. Thus, the existence of quadruple, octuple etc. static moments follows directly from the structure of the nucleus. The quantitative side of the problem is caused by several nuclear peculiarities. As the value of the quadruple moment is equal to the electric charge (q) multiplied by the product of dipole basis and the distance between the centers of dipoles, *the first problem* concerns the inequality of negative wave charges to a free negative electric charge. The effectiveness of a negative wave as negative charge must depend on the bond between the neutron and its neighbors. *The second problem* concerns the changes of the distances between neighboring proton and neutron (the basis of a dipole) at oscillation. As the intensity of oscillation depends on the number of nucleons, the quadruple moment has to depend on the "magic" number of the nucleus. *The third problem* concerns the influence of the spin and nuclear rotation. As the rotation depends on the spin and the spin depends on the oscillation, it becomes difficult to separate the cause from the effect. So, the absence of quadruple moment in nuclei with spin ½ η is probably due to the absence of a perceptible nucleon oscillation. *The fourth problem* concerns the way of quadruple moment compensation. Evidently the quadruple moment of a nucleus is the result of the balance of the quadruple moment of the nuclear branches. The absence of quadruple moment in even/even nuclei can be explained only as a result of full compensation. Probably the quadruple moment of each branch is a sum of the separate quadruple moments, but it is also possible that a generalized combination of the charges along the branches forms a common quadruple moment of each branch.

Anyway, the quadruple moment (Q) of the nucleus ought to be proportional to the number of nucleons (A) due to a generalized influence of the distance between the dipoles, and to $A^{0,5}$ - due to the influence of dipole basis (distance between charges) on the period of nucleon oscillation, or

$$Q = kA^{3/2} + b$$

where k and b are coefficients.

The deviation from this dependence is caused by the spin and "magic" of the nucleus.

3.4.4. Nuclear energy levels

The nucleus is a complex system of particles and forces. The strong force strives to keep nucleons in the lowest possible energy level, which corresponds to a bond between nucleons with opposite spins. The repulsive forces (electric and magnetic) relax the ties to certain elevated energy level. So the energy of a nucleus in the basic state consists of the energy of nucleon oscillation and the energy of de Broglie waves (or spin energy of the extra nucleons). Energy from outside influences the ties between nucleons by causing spin changes – from an "opposite spin" to a "parallel spin". But the change of one bond reflects on all remaining ties in the nucleus and this dependence determines the energy necessary for the next level of excitation. This is why the first energy level in the even/even, and especially in magic nuclei, is bigger than in the rest of the nuclei and is in the range of one to several MeV. The first energy level in nuclei having odd number of nucleons is usually on the order of KeV, and is connected to the spin increase of an extra nucleon. But the next level, in the range of MeV, is connected to destabilization of the nucleus as a whole. The energy necessary for subsequent levels of excitation gradually decreases, because it concerns smaller groups of nucleons and ties, which have been relaxed in some degree at the lower level of excitation. When the energy of excitation is equal to 7 – 8 MeV, the nuclear energy spectrum becomes *continuous* – practically all nucleons become loosely bonded. The nucleons oscillation becomes independent or *chaotic*. The energy levels over 12 to 14 MeV is the range of *Gigantic resonance*. Depending on the character of the force of excitation, the nucleons can take part in different types of collective motion. The mode of nucleons oscillation determines the type of the gigantic resonance. For instance, protons and neutrons in a branch oscillate in opposite directions in the *dipole gigantic resonance*. In a *monopole gigantic resonance,* protons and neutrons in one branch oscillate together (in one direction) but oppositely to the direction of oscillation in the other branch. The possible deformation of the nucleus (deviation from the straight line arrangement of nucleons) causes formation of other resonances, for instance *quadruple gigantic resonance*.

Without doubt it is the structure of the nucleus that predetermines the spectrum of energy levels. A nuclear spectrum calculation will be possible when *the potential* (the binding energy dependence on the distance) of the nucleon interaction will be determined. The way of nucleon arrangement in the nucleus predetermines the importance of *two particles potential* as a fundamental one, because *the strong interaction in a nucleus consists of interactions between two nucleons.* Evidently the two particle potential depends on the type of nucleons (the charge independence is an illusion). The potential between proton and neutron cannot be the same as between two neutrons. Moreover the potential must depend on: 1. The nucleon spin orientation; 2. The type and the spin orientation of neighboring nucleons; 3. The proton and the neutron stress; 4. The type of nucleon numbers (odd, even); 5. The "magic" of numbers, etc. The task is not very easy, but not as difficult as the problem of the many body interaction. Anyway, the problem is soluble.

3.5. STRUCTURE AND PROPERTIES OF ATOMIC NUCLEI

The properties of nuclei used in this book are taken from the *Nemetz and Hofman's Reference Book of Nuclear Physics [1]* . At first, for schematic presentation arrows will be used showing the spin direction: $\uparrow$ - for proton, $\Uparrow$ -for neutron.

The values of the nuclear and nucleon spin are expressed in η units.

The values of magnetic moment are expressed in nuclear magnetons.

3.5.1. Hydrogen

The protium, the ordinary hydrogen nucleus (H_1^1) is "structureless" because it consists of only one proton.

The deuteron (H_1^2 or d) is a stable isotope, consisting of one proton and one neutron, with following properties: Binding energy $E_b = 2,22$ MeV; Spin J= η; Magnetic moment $\mu = + 0,86$ nuclear magnetons; Quadruple moment Q = +0,0028 barns related to the proton charge. These properties testify for the same orientation of the spins of the nucleons. Schematically the deuteron can be represented as ↑↑↑ . One of the reasons for the parallel spins of the proton and neutron in deuterium is the general rule for nuclear structure. The nucleus of the deuterium consists of two nucleons, and each of them represents a separate branch. The spin directions of protons in the left branch must be parallel to the spin direction of neutrons in the right branch, and hence the spin of deuterium must be equal to η. The other cause could be the interaction between the waves. The parallel spins provide possibility for a stronger bond of the negative wave – practically it is tied with two positive waves. This is the cause for the neutron stability despite of the small binding energy. The relatively small binding energy testifies for a small degree of matter- waves interference and as a result the magnetic moment is practically equal to the difference between the magnetic moments of the nucleons. In addition, a deuteron nucleus with opposite spins cannot exist because spin zero is incompatible with existence of a magnetic moment. A proton and a neutron cannot form a Cooper pair. The cause for the existence of quadruple electric momentum is a puzzle. Probably the negative wave of the neutron plays the role of a negative charge between two positive charges and it acts like a quadruple...

The triton (H_1^3 or *t*) consists of one proton and two neutrons and has the following properties: E_b = 8,48 MeV, J = ½ η; μ = +2,98; Q = 0,0028. The triton is radioactive. It undergoes β^- decay with a half-life of 12,26 years. It is evident from the values of spin and magnetic moment that the two neutrons have opposite spins. The most probable schematic of triton is ⇓↑↑↑ . The bond of the negative wave of the left side neutron with the proton is not firm enough to prevent the trend towards straightening. β^- decay of this neutron occurs together with charge exchange between the right side neutron and the proton to form the nucleus of He_2^3 (↓↑↑↑). The spin of the triton is set by the proton. Because of the spin pulses the nucleus rotates as a whole in direction opposite of the spin. This rotation increases the magnetic moment of the right side neutron with approximately 0,2 in order to reach the value of + 2,98.

According to Evans [9], two more isotopes of hydrogen exist:

H_1^4 is a short life time isotope which decays via the release of a neutron. Its most probable schematic is ⇓↑↑↑↑ . The neutron at the end of the right side is loosely tied and easily separates. According to this schematic $J = \eta$ and $\mu = +0,7$ to $+0,8$ (approximately).

H_1^5 is β^- radioactive isotope with a half-live of approximately 0,1 s. Its most probable schematic is ⇃⇃↿↾↾ with main properties $J = \frac{1}{2}$ η and $\mu \approx +2,8$. The neutron drip-line passes via H_1^5. It is impossible for one proton to keep more than four neutrons.

3.5.2. Helium

He_2^3 is a stable isotope with properties: $E_b = 7,72$ MeV; $J = \frac{1}{2}$ η; $\mu = 2,13$. Its schematic of nucleon arrangement is ↑↿⇂. The cause for the relatively small binding energy is the large area of electrostatic repulsion between the protons (see Fig. 5). Because of the nuclear rotation, the magnetic moment of the right side proton becomes bigger with approximately 0,20 to form the total magnetic moment of 2,13. At temperature near $0°K$ two nuclei form Cooper pair (↑↿⇂↑⇂⇂ with $J = 0$; $\mu = 0$) and the gas He_2^3 transitions into a state of superfluidity.

He_2^4 is the most stable nuclide in nature and is considered as one particle (α- particle) with properties: $E_b = 28.30$ MeV; $J = 0$ and $\mu = 0$. The schematic of arrangement is ↑⇂↿⇂. The opposite spin direction and the increased distance between protons (screened by two neutrons) are the factors for a large binding energy. The binding energy of He_2^4 will be discussed in more detail in Chapter 3.

He_2^5 is an unstable nuclide and decays to He_2^4 and neutron (n) in a time less than 10^{-8} s. It has a spin $J = 3/2$ η and a binding energy of 27,41 MeV – less than the binding energy of He_2^4 . Probably the structure of He_2^5 is ⇂↑ $\overset{⇑}{—}$ ⇂↿ . The neutron in the middle is displaced in "orbit". Between the two nucleons on the left side a spin and a charge exchange take place, which leads to ↑⇂↿⇂↿ , and to separation of the outer (right side) neutron. The result is just the nuclide He_2^4 .

He_2^6 has a binding energy of 29,27 MeV, a spin $J = 0$, magnetic moment $\mu < 0,16$ and undergoes a β^- decay. Its most probable schematic is ⇃↑⇂↿⇂↿ . The neutrons at both ends are loosely tied –

by only 0,5 MeV each. The β^- decay occurs in one of the neutrons in the middle and goes along with a charge exchange between the nucleons of both sides.

He_2^8 is an unstable nuclide with E_b =31,36 MeV; J = 0; μ = 0. The β^- decay goes along with a neutron release. The neutron drip-line passes via He_2^8 because all places for neutrons are filled up. Its schematic is ⇊⇊↑⇊⇈↓⇈⇈. In the β^- decay one of the outside neutrons separates, the neutron on the other end transforms into a proton. Then a charge and a spin direction exchanges follow, which leads to the formation of the nuclide Li_3^7. The binding energy release is approximately 8 MeV.

3.5.3. Lithium

Li_3^5 is an unstable nuclide. Its schematic of arrangement is ↑⇊↑⇈↓. According to this schematic, the spin is ½ η (due to the proton in the center) and the magnetic moment must be near to + 2,8. Probably the central proton undergos β^+ -decay. Between the neutron and proton on the right side occure charge and spin exchanges. The schematic becomes ↑⇊⇈↓⇈ and the neutron on the right side separates.

Li_3^6 is a stable nuclide, having E_b = 31,99 MeV; J = η; μ = +0,82; Q = 0,0008. Its schematic of arrangement is ↑⇊↑⇈⇈↓. The value of the spin is due to the middle proton-neutron pair having parallel spins, and the magnetic moment is approximately equal to the difference of magnetic moments of these two nucleons.

Li_3^7 is a stable nuclide. E_b = 39,25; J = 3/2 η; μ = +3,26; Q = - 0,04. Its schematic is ↑⇊⇊ ↑̅ ⇈⇈↓. The extra proton in the middle is

displaced into "orbit". The nuclear spin consists of ½ η spin of the extra proton and 1 η- spin of its de Broglie wave ("orbit"). The total "proton-orbital" spin becomes 3/2 η. The magnetic moment (+3,26) also consists of extra proton magnetic moment (approximately + 2,8) and the magnetic moment of de Broglie wave ("orbit") estimated to approximately +(0,4 – 0,5) of each η- unit. For instance at spin of 7/2 there are 3 η- units and the magnetic moment of "orbit" must be approximately +(1,2 – 1,5). From now on the sum of spins and the sum of magnetic moments of the extra nucleon and de Broglie wave will be interpreted as a spin and magnetic moment of the extra nucleon.

Li_3^8 undergoes β^- decay and separation in 2α particles (half-life 0,84 sec). Its schematic is ↑⇓⇓ $\overset{↑}{—}$ ⇑⇑↓⇑. The spin J = 2 η consists of 3/2 η spin of the extra proton in the center and ½ η spin of the extra neutron (on the right side of the extra proton). The magnetic moment of the nucleus (μ =+1,65) consists of the magnetic moment of the extra proton (approximately +3,26 - see Li_3^7) and the magnetic moment of the extra neutron, estimated at approximately -1,60. The extra neutron undergoes a β^- decay and after several spin and charge exchanges the nucleus separates in 2α particles releasing binding energy of approximately 15 MeV.

Li_3^{11} is the last unstable nuclide with a schematic ⇓⇓↑⇓⇓↑⇑⇑↓⇑⇑. All places for neutrons are occupied and hence this nuclide is on the neutron drip-line.

3.5.4. Beryllium

Be_4^7 is an unstable nuclide. An electron capture and release of γ quanta occurs (half-life 53,3 days). Its schematic is ↑⇓↑ $\overset{⇑}{—}$ ↓⇑↓. The spin (3/2 η) and the magnetic moment are due to the extra neutron in the middle. One of the two inside protons captures an electron and after charge and spin exchanges the nucleus turns into Li_3^7.

Be_4^8 is an unstable nuclide and separates in 2α particles in 10^{-16} s. Its schematic is ↑⇓↑⇓⇑↓⇑↓. Before separation only a charge exchange between the two inner proton-neutron pairs occurs.

Be_4^9 is a stable nuclide. E_b = 58,17 MeV; J = 3/2η; μ = -1,18; Q=+0,032. Its schematic is ↑⇓⇓↑ $\overset{⇑}{—}$ ↓⇑⇑↓. The spin is due to the spin of the extra neutron in the middle. The magnetic moment is a balance between the magnetic moment of a branch consisting of 3 neutrons (on the left) against the magnetic moment of a branch with 2 neutrons (on the right side). The number 3 is near the magic number 4, which means a reduced (to approximately – 1,60) magnetic moment of the neutrons in that branch against the approximately -1,80 magnetic moment of neutrons in the branch with 2 neutrons. The balance will be 3. (-1,60) – 2(-1,80) $\approx$ -1,20.

Be_4^{10} is a long living isotope with a half-life of $2,6.10^6$ years. The schematic is ↑⇊↑⇑↓⇑↓. One of the central neutrons undergoes a β^- decay and after charge and spin exchanges the nucleus turns into the nuclide B_5^{10} .

Be_4^{11} and Be_4^{12} are β^- radioactive nuclides with schematics of nucleon arrangement ⇊↑⇊↑⇑↓⇑↓ and ⇊↑⇊↑⇑↓⇑↓⇑ respectively.

Be_4^{14} is the last possible of the Beryllium neutron rich nuclides, which stands on the neutron drip-line (although the neutron separation occurs at Be_4^{12}).

The way of the nucleon arrangement representation by arrows is visual, but becomes inconvenient with increasing number of nucleons. It is reasonable to mark the protons by the letter **p** , and neutrons by the letter **n**. So the schematic of arrangement of the nucleons in Be_4^{14} ⇊↑⇊↑⇑↓⇑↓⇑ can be indicated as **nnpnnpn:npnnpnn**.

For an even shorter record, the nucleons can be represented in groups and the number of the groups can be marked by the number before the round brackets. So the structure of the nucleon Be_4^{14} can be written as **2(nnp)n:n2(pnn)**. The colon in the middle indicates the border between the two branches of nucleons.

3.5.5. Boron

For simplicity, further in the book the properties of nuclides will be given in brackets in the following order: mode of decay, half-life, binding energy, spin, magnetic moment and quadruple moment.

B_5^8 ($\beta^+,2\alpha$; 0,77 sec; E_b = 37,74 MeV; J = 2 η). Its schematic of nucleon arrangement is **pnpn:ppnp** . (It is not difficult to keep in mind that in the left branch the proton spins direction is herein referred to as "from us", opposite to the direction of neutron spins. The direction of spins in the right branch are opposite to the spin directions in the left branch). The spin of B_5^8 is a sum of extra nucleon spins – proton spin 3/2 η, and neutron spin 1/2 η. The inside proton of the right branch undergoes β^+ decay and after charge and spin exchanges the nucleus separates in two α particles.

B_5^9 (p + 2α; E_b = 56,32 MeV). The schematic is **pnpnp:npnp**. The probable value of the spin should be 3/2 η, and magnetic moment approximately +2,70, - determined by the extra proton in the center. The proton stress and the trend for α - particles formation tears the nucleus along the middle and the loosely tied extra proton is set free.

B_5^{10} (stable; E_b = 64,75 MeV; J = 3; μ = +1,80; Q = +0,074), **pnpnp:nnpnp**. The nuclear spin is a sum of extra proton spin (5/2 η) and extra neutron spins (½ η). The difference between the magnetic moment of the central proton and neutron must be approximately 0,80 (parallel spins). The magnetic moment of the "orbit" with a spin of 5/2 η is equal to approx. 2. 0,50 = 1,00 and hence the total sum will be 1,80.

B_5^{11} (stable; E_b = 76,21 MeV; J =3/2; μ =+2,69; Q = 0,036), **pnpnnp:nnpnp**. In this nuclide the influence of the "magic" of numbers becomes noticeable. At absance of "magic" the magnetic moment of the nucleus must be approximately 3.30 (0,5 of the "orbit" and 2,8 of the extra proton). The difference of approx. 0,70 is because the magnetic moment balance of two proton rows – the magnetic moment of each proton in the left row is less than the magnetic moment of each proton in the right row because the number 3 is nearer to the magic number 4 than the number 2 is. The "magic" of numbers reduces the magnetic moment of all nucleons in the row.

B_5^{12} (β^-,α,γ; 20,3 ms; E_b = 79,58 MeV; J = 1; μ = +1,00). The schematic is **pnpnnp:nnpnnp** . The spin is a sum of the extra proton and extra neutron spins – each equal to ½ η. The sum of their magnetic moments is bigger than 0,80 (parallel spins), because the neutron is in the "magic number" (4) row. The second neutron in the center undergoes a β^- decay and after charge and spin exchanges the nuclide decays in three α -particles.

The neutron drip-line ought to pass through B_5^{17} - **nnpnnpnnp:nnpnnpnn**.

3.5.6. Carbon

C_6^9 (β^+, p, 2α; 0,127 s; E_b = 39 MeV). The schematic is **pnppn:ppnp.** The left endmost proton "drops". Its neighbor pairs **nppn** exchange the charges to become **pnnp.** In the other **ppnp** branch the proton inside undergoes a β^+ decay and so two α-particles are formed. Probably the proton drip-line ought to pass through C_6^9.

C_6^{10} (β^+, γ; 19,48 s; E_b = 60,32 MeV; J = 0), **pnpnp:npppnp.** The first proton in the couple **pp** of the right branch undergoes β^+ decay.

C_6^{11} (β^+; 20,4 min; E_b = 73,44; J = 3/2; μ = 1,03; Q = 0,031). The schematic is **pnpnpn:pnpnp.** The magnetic moment of even/odd nuclei ought to be at least negative. The anomaly (μ = 1,03) is due to the proximity of neutron numbers 3 in the left row to the magic number 4. Evidently, the sum of magnetic moments in the row with 3 neutrons is less than that sum in the right row containing 2 neutrons only. Probably the balance of the magnetic moments looks like: 3.(-0,95) − 2.(- 1,9) = + 1.00. One of the central protons undergoes β^+ decay. Charge and spin exchanges occur at the B_5^{11} formation.

C_6^{12} (Stable; E_b = 92,16 MeV; J = 0), **pnpnpn:npnpnp.**

C_6^{13} (Stable; E_b = 97,11 MeV; J = 1/2; μ = +0,70), **pnpnnpn:npnpnp** . Similarly to C_6^{11}, the magnetic moment is positive because the number of the neutrons in the left row is 4 (magic number). Probably the magnetic moment of neutrons in this magic row does not exceed − 0,55; [-0,55 x 4 −(-0,95 x3)] = + 0,65 .

C_6^{14} (β^-; 5730 years; E_b = 105,29 MeV; J = 0), **pnpnnpn:npnnpnp.** One neutron of the couple **nn** undergoes a β^- decay.

C_6^{15} (β^-, γ; 2,49 s; E_b = 106,50 MeV; J = ½). The schematic is **pnnpnnpn:npnnpnp.**

The neutron drip-line ought to pass through C_6^{20} - **nnpnnpnnpn:npnnpnnpnn**

3.5.7. Nitrogen

N_7^{12} ($\beta^+, 3\alpha, \gamma$; 11 ms; $E_b = 74,00$ MeV; J $= 1$; $\mu = 0,46$), **pnpnpp:npnpnp**. The magnetic moment is less than the usual ($\approx 0,8$) for the odd/odd nuclei, because the extra proton is in a row with the magic number of protons. The extra proton undergoes a β^+ decay and by exchanging of charge and spin directions the nucleus divides into 3α particles.

N_7^{13} (β^+; 9,97 min ; $E_b = 94,11$; J $= 1/2$; $\mu = 0,32$), **pnpnpnp:npnpnp**. The magnetic moment is too small for a normal odd/even nuclide. The cause is the same as above at N_7^{12} - the magnetic moment of protons in the left, "magic" row is less than the magnetic moment of protons in the right row having 3 protons.

N_7^{14} (Stable; $E_b = 104,66$ MeV; J $= 1$; $\mu = 0,40$; Q $= +0,01$), **pnpnpnp:nnpnpnp**. The analogy between N_7^{14} and N_7^{12} is evident.

N_7^{15} (Stable; $E_b = 115,49$ MeV; J $= 1/2$; $\mu = -0,28$), **pnpnpnnp:nnpnpnp**. The anomaly of the magnetic moment is bigger than in N_7^{13}, probably because the branch of the extra proton is double magic (Z $= 4$; N $= 4$), and the sum of the magnetic moment of 4 protons in the left row is smaller than that sum in the right row (containing 3 protons).

N_7^{16} (β^-, γ; 7,14 s; $E_b = 117,98$; J $= 2$); **pnpnpnnp:nnpnnpnp**. The nuclear spin is a sum of the spins of the extra nucleons: $J_p = 3/2, J_n = 1/2$.

The neutron drip-line ought to pass through N_7^{23} - **nn3(pnn)p:3(nnp)nn.**

3.5.8. Oxygen

O_8^{13} (β^+, p; 0,7 ms). The probable schematic is **ppnpnpn:ppnpnp**. The endmost proton of the left branch separates. One of the protons in the couple **pp** in the right branch undergoes a β^+ decay.

O_8^{15} (β^+; 123 s; J = ½ η; μ = 0,72), **4(pn):p3(np)**. The positive magnetic moment is anomalous for an even/odd nuclide. The reason is the belonging of the extra neutron to a doubly magic branch.

O_8^{16} (stable; J = 0; μ = 0), **4(pn):4(np)**, double magic branches.

O_8^{17} (stable; J =5/2; μ = -1,89), **2(pn)pnnpn:4(np)**. The magnetic moment is a balance between the magnetic moment of neutrons in the right row containing 5 **n** and in the left magic row.

The neutron drip-line ought to pass through O_8^{28}, **nn4(pnn):4(nnp)nn**.

3.5.9. Fluorine

F_9^{17} (β^+, ε; 109,8 min; J = 5/2; μ = + 4,72), **4(pn)p:4(np)**. The magnetic moment is a balance between the μ of a row containing 5 protons, μ of a magic row (4 protons), and μ of the extra proton "orbit" (with spin 5/2 η).

F_9^{19} (stable; J = ½ ; μ = + 2,67), **4(pn)np:n4(np)**. The schematic is analogous to the schematic of F_9^{17} (only two neutrons more). A smaller spin (½ η) of the extra proton predetermines a smaller magnetic moment.

F_9^{20} (β^-, γ; 1,56 s; J = 2; μ = +2,09), **2(pn)pnnpnp:nnpnpnnpnp**. The spin consists of the extra proton spin (½η, accepted by an analogy with the spin of F_9^{19}) and an extra neutron spin (3/2 η). The μ is a balance between the magnetic moment of two proton rows (ratio 5/4) and magnetic moment of two neutron rows (ratio 6/5).

3.5.10. Neon

Ne_{10}^{19} (β^+; 17,4 s; J = ½ ; μ = - 1,88), **4(pn)pn:p4(np)**. The μ-balance is between the magnetic moment of neutrons in rows with neutron number ratio 5/4.

Ne_{10}^{21} (stable; J = 3/2; μ = - 0,66), **4(pn)npn:5(np)** . Probably the spin 3/2η is related to the proton on the left of the colon. The positive μ of the "orbit" influences the balance of magnetic moment of the neutron rows (ratio 6/5).

3.5.11. Regularities in nucleon arrangement

The adduced examples of nucleon arrangement are sufficient to draw the following conclusions:

- The basic structural unit in the low mass nuclei is the couple **pn.** With increasing of the mass and the ratio **n/p**, a gradual replacement of the couple **pn** by a group **pnn** occurs. The existence of 3 neutrons with equal spin orientation **nnn** is forbidden. Probably the configuration of 3 neutrons with a different spin orientation **nn:n** can exist only in the middle of the heaviest short lived neutron rich isotopes.

- All even/even nuclei have a **pn:np** unit in the middle. The rest of the **pn** couples and **pnn** groups are symmetrically arranged in both branches.

- The even/odd nuclei have the same **pn:np** unit in the middle, but also a **pnn** on one side instead of **pn**. In the light proton rich nuclei the middle unit is **p:np**.

- The odd/odd nuclei have **p:n** unit in the middle.

- The odd/even nuclei have **np:n** unit in the middle.

3.5.12. Calculation of nuclear structure

The nuclear structure is the way of nucleon arrangement. Each nucleus consists of a central group of nucleons and the nucleons standing in the branches, distributed in groups **pn** and **pnn**. The number of **pnn** groups is equal to the difference N − Z. The way of arrangement of groups in the nuclei is subordinate to the necessity to space out the protons in a way to reduce the proton stress. Hence, the arrangement of the **pnn** groups must begin from the center where the proton stress is at maximum. In the heavy mass nuclei several zones can be formed: a central zone with a **pnnpnn...** structure; two periphery zones with a **pnpn...** structure, and 2 transitional zones with a **pnnpnpnpnn...** structure. Of course this way of arrangement is only logically the most probable one. In each multi-particle system the less probable variants are also possible

Several examples of multi-nucleon structures are shown bellow:

Even/even nuclei

1. Ca_{20}^{40} - **10(pn):10(np)**; Ca_{20}^{46} -
3(pn)3(pnpnn)pn:np3(nnpnp)3(np).

2. Sn_{50}^{124} -

6(pn)3(pnpnn)3(pnnpnpnn)3(pnn)pn:np3(nnp)3(nnpnpnnp)3(nnp np)6(np)

Because of the symmetry in the structure of even/even nuclei the right branch is obvious and will not be written explicitly.

3. Pb_{82}^{208} - **4(pn)10(pnpnn)4(pnnpnpnn)4(pnn)pn:np4(nnp)…**

4. U_{92}^{234} - **18(pnpnn)2(pnnpnpnn)3(pnn)pn:np3(nnp)…**

U_{92}^{236} - **16(pnpnn)3(pnnpnpnn)4(pnn)pn:np4(nnp)…**

U_{92}^{238} - **14(pnpnn)4(pnnpnpnn)5(pnn)pn:np5(nnp)…**

Even/odd nuclei consist of the branches of the two neighboring even/even nuclei. For instance the nuclide U_{92}^{235} consists of one branch of U_{92}^{236} and one branch of U_{92}^{234} or

16(pnpnn)3(pnnpnpnn)4(pnn)pn:np3(nnp)2(nnpnpnnp)18(nnpnp).

Odd/odd nuclei. The structures of two long-living nuclides are given:

1. K_{19}^{40} - **8(pn)pnnp:2(nnp)7(np).**

The left branch is equivalent to a branch of Ca_{20}^{40} although the structure is slightly modified – the places of the last neutron and proton are exchanged (**pn** → **np**). The right branch is just equal to a branch of Ar_{18}^{40}.

2. Nb_{41}^{94} - **10(pn)4(pnpnn)2(pnn)p:3(nnp)4(nnpnp)9(np).**

The left branch is equivalent to a branch of Mo_{42}^{94}, the right one – to a branch of Zr_{40}^{94}.

Odd/even nuclei

1. Nb_{41}^{93} - **10(pn)4(pnpnn)2(pnn)p:2(nnp)4(nnpnp)10(np).**

The two branches are symmetric in relation to the central proton.

2. Bi_{83}^{209} - **5(pn)10(pnpnn)4(pnnpnpnn)4(pnn)p:4(nnp)…**

3.5.13. Anomaly in nuclear properties

From the examination of the nuclear properties in [1] the following remarks can be drawn:

A high nuclear spin is typical for nuclei having numbers of nucleons immediately after the magic numbers, as for instance $O_8^{17}, O_8^{19}, (J = 5/2\eta); F_9^{17}, F_9^{21}, (J = 5/2\eta); S_{16}^{37}, (7/2\eta);$

$Ca_{20}^{41}, Ca_{20}^{43}, Ca_{20}^{45}(7/2\eta);$ All isotopes of Scandium $(7/2\eta);$ The odd/even isotopes of Indium $(9/2\eta);$ The odd/even isotopes of Bismuth $(9/2\eta)$ etc. If you take the value of the spin as a criterion, the "magic" of Cadmium seems equivalent to the "magic" of Tin. Because of the superposition the influence of "magic" is spread (in lesser degree) over many other numbers, as for instance Number 24, (3x8) – the cause for the elevated spin of isotopes of Manganese (5/2 and $7/2\eta$); Number 40, (2x20) – the cause for the high spin of odd/even isotopes of Niobium $(9/2\eta)$ etc.

These results are in agreement with the assumption that a high spin is a result of a big difference in the intensity of nucleon oscillation in both rows.

As a balance between the magnetic moments of the two branches, the magnetic moment of the nucleus strongly depends on the "magic" of numbers. The odd/even isotopes of elements, which are next to the elements having "magic" number of protons, must possess a big positive magnetic moment, for instance $F_9^{17}(+4,62), F_9^{19}(+2,62);$

$Sc_{21}^{43}(+4,62), Sc_{21}^{47}(+5,34);$ $Nb_{41}^{93}(+6,17); In_{49}^{109-115}(+5,53);$

$Bi_{83}^{203-209}(+4,60 - +4,10).$ All these nuclides possess a high spin (see above). Thus the role of the magnetic moment of the de Broglie wave ("orbit") becomes clear.

The magnetic moment of the nucleus is very often a puzzle, because the components of its balance are not clear enough. As a rule, the proximity of a nucleon number to the magic one creates an "anomaly" of the nuclear magnetic moment. Such anomaly has been noticed for the magic number 8 (Oxygen). The next anomaly must be at number 16 (2x8). As a matter of fact S_{16}^{33} at a J=3/2 has a positive magnetic moment +0,64 because the extra neutron is in a row containing 9 neutrons (close to the magic 10). Cl_{17}^{35} - with a spin of the extra proton (J = 3/2) has an anomalously small $\mu = +0,82$, because the proton ratio in the rows is 9/8. Ar_{18}^{37} - with a spin 3/2 has a positive μ + 0,95, because the extra neutron is in a row with a magic number 10 of neutrons.

Another area of anomaly in magnetic moments is found in Cadmium. The four odd/even nuclides of Silver (the element before Cadmium) have small (½) spins and small negative μ (- 0,10 to – 0,15), because the extra proton is in a magic row 24 **p.** The four odd/even nuclides of Indium (the element after Cadmium and before Tin) have a big 9/2 spin and a big μ (+5,5), because the extra proton is in a row with 25 protons. The 4 odd/even nuclides of Antimony (the element after Tin) have a spin 5/2 (smaller than the spin 9/2 of Indium) and a μ (+2,67 - +3,45). Hence Antimony is farther from the "magic" number than Indium.

The exactness of magnetic value estimation quickly decreases with the nucleus number increase. The problem lies in the increased number of participants (nucleons) in the balance with decreased certainty in the values. The influence of the "magic" of numbers gradually decreases with nucleon number increasing, and becomes unequal for all nucleons in a row. The influence of the nuclear rotation on the magnetic moment of each nucleon gradually decreases, but as the number of nucleons increases, the sum can become significant. It is possible that some changes in the rules of addition, shielding and compensation of nucleon magnetic moments (as wave processes) can occur in the middle and heavy mass nuclei. Of course the solution of all these problems is a task for the future.

In the Reference book [1], there are some real anomalies:

a) Odd/odd nuclei with a spin zero and magnetic moment zero.

The problem of spin compensation in odd/odd nuclei is similar to the problem of absence of the nucleus H_1^2 with spin zero. It is evident, that simultaneous compensation of the spin and the magnetic moment between a proton and a neutron is impossible. Then, how to explain the existence of Cl_{17}^{34}; V_{23}^{46}; Ga_{31}^{66}; Br_{35}^{78}; Ho_{67}^{166}; Lu_{71}^{170} etc.? Each of them has a spin zero and a magnetic moment zero – a "miraculous" compensation of spins and magnetic moments of the extra proton and extra neutron. Evidently besides the extra proton and extra neutron the compensation includes another nucleon. For instance, the nuclide Br_{35}^{80} (J = 1; μ = +0,52) is a normal one and in comparison with Br_{35}^{78}

has only 2 more neutrons with opposite spins and magnetic moments. Hence these 2 neutrons cannot have a noticeable influence on the interaction between the extra proton and extra neutron, but they can cause spin changes in the neighboring nucleons. Thus, if a proton in a right branch obtains spin 3/2, the spin of "orbit" (η) can compensate the spin of the extra nucleons, and the magnetic moment of "orbit" ($\approx$ -0,5) ought to compensate for the difference between magnetic moments of extra nucleons. Of course, the influence of the "magic" of numbers ought to be taken into consideration in the balance of magnetic moments.

b) *Even/even nuclides with a spin zero which possess magnetic moment*

Some even/even nuclides of rare earth elements having nuclear spin zero, have also magnetic moment, for instance, Sm_{62}^{152} ($\mu = 0,56$); Sm_{62}^{154} ($\mu = +0,42$); Gd_{64}^{154} ($\mu = 0,73$); Gd_{64}^{156} ($\mu = 0,70$).

It is because a proton and a neutron (but not a pair **pn**) are in "orbits" with opposite direction of rotation. The total spin remains zero, but the balance between the μ of the "orbits" determines a μ of the nucleus.

3.5.14. Anomaly in Electric quadruple moment

The analysis of the values of *nuclear electric quadruple moment* taken from [1] shows a general trend towards its increasing with the increase of nucleon number A. The biggest part of the positive values are in agreement with the dependence: $\lg Q = 3/2 \lg A - 0,30$. The main causes for the deviations are:

1. *The magic numbers of nucleons.* The even/odd nuclides with magic number of protons have a small positive or negative quadruple moment ($O_8^{17}, S_{16}^{33}, Sn_{50}^{121}, Nd_{60}^{143}$, etc). Some of odd/even nuclides with magic neutron number have also negative quadruple moment [V_{23}^{51} ($N = 28$), Pr_{59}^{141} ($N = 82$)]. The odd/even (and some odd/odd) nuclides with proton number next to the magic numbers have a negative quadruple moment ($Li_3^7, Cl_{17}^{35}, Sc_{21}^{43,47}, Cu_{29}^{63,65}, Sb_{51}^{115}, Bi_{83}^{203}$, etc). Most of these nuclides have a high spin and a big magnetic moment. Evidently the spin and magnetic moment influence the balance between the quadruple moments of nuclear branches.

2. *The distance from the magic numbers.* The nuclides in the region between the magic numbers have increased quadruple moment.

This can be noticed in the middle of the range between proton numbers 50 and 82 , in the region of numbers 57-71 (rare earth), $Er_{68}^{163}\,(Q=+3,9), Er_{68}^{167}\,(+2,83)$, $\qquad Tm_{69}^{166} \qquad (+4,50)$, $Lu_{71}^{175}\,(+5,60), Lu_{71}^{176}\,(+8,00)$ etc. and after the proton number 82 – the range of the Actinides $Ac_{89}^{227}\,(+1,70)$, $Th_{90}^{229}\,(+4,60)$, $U_{92}^{235}\,(+4,1)$, $Pu_{94}^{241}\,(+5,60)$, etc. Evidently in the middle of the diapason between the magic numbers the nucleon oscillation is more vigorous and the distances between the nucleons become bigger. This is equivalent to increased bases of the dipoles and to increased distance between the dipoles. The result is an increased quadruple moment as a product on the basis of dipoles and the distances between them.

CHAPTER 4

ANALYSIS OF NUCLEAR REGULARITIES

4.1   NEUTRON / PROTON RATIO

The number of protons Z gives the name of the nuclides, sets the number of electrons in atoms, and determines the limits of the number of neutrons N – (the neutron drip line). The role of neutrons in nuclei is to shield and simultaneously bond the protons. In light stable nuclides one neutron between two protons is enough for shielding and Z and N can be practically equal. With increasing Z, the increased proton stress needs extended distances between protons (or thicker shielding) and leads to doubling the neutron shield. In the heavy stable nuclides the ratio N/Z does not exceed 1,5. Of course the number of neutrons depends on the type of proton number (even or odd, close or far from the "magic" numbers), but a general trend can be seen on Fig. 15, where the number of the excess neutrons $\Delta N = N - Z$ is plotted as a function of the number of protons for all stable nuclides. The neutrons are located in a large strip along the line $\Delta N = 0,0065Z^2$. At a close examination of Fig. 15 it becomes clear, that the sections with equal follow sections of rapid ΔN increase. Evidently, the "magic" proton and neutron numbers play a definite role in the successions of the nucleons. In general, however, each nucleus is individual.

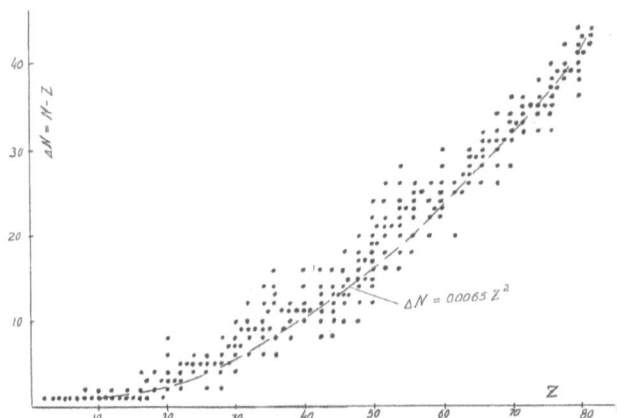

Fig. 15. Excess of neutron (ΔN) dependence on the proton number (Z)

In Fig. 16 the number N is plotted as a function of the number Z. The Stable isotopes are disposed in a strip (2) close to line $N = Z + 0,0065Z^2$. This line separates the nuclides which undergo β^- decay from those which undergo β^+ decay. The isotopes, which could be in possible use (because of its noticeable half-life) are disposed in the hetched area (3). All theoretically possible isotopes are disposed between neutron drip-line (1) and proton drip-line (4).

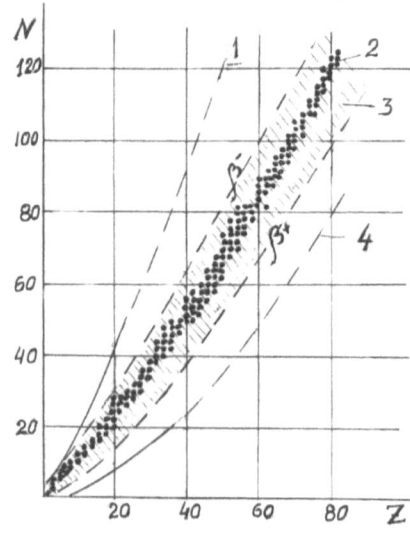

Fig. 16. Neutron-proton relation
1. Neutron drip-line; 2.) Strip of stable nuclide disposition;
3. Strip of isotopes of possible use; 4. Proton drip-line

- *The neutron drip-line* in the light nuclides is related to the filling up all possible places (*neutron saturation*) – two neutrons between the neighboring protons and two neutrons on each side. Thus the number of neutrons becomes N = 2 Z + 2. But probably in heavy mass nuclei, because of proton stress increasing, a short life time formation of **nnn** groups could be possible. Hence in the heaviest nuclei, probably the neutron drip-line ought to approach the line N ≈ 3 Z. The estimated drip-lines of middle and heavy mass nuclides are given by broken lines in fig 16.

- As the repulsion between protons is stronger than repulsion between neutrons, a precise location of proton drip-line is difficult, or even impossible, as the speed of proton release because of proton stress could be comparable with "dripping" caused of proton saturation. By analogy with neutron dripping it reasonable to suppose, that at light mass nuclei the ratio neutron/proton ought to be near two (N/Z ≈ 2) and will rapidly decrease with protons number increasing

In the area between the two drip- lines (spanning from Z=1 to Z= 83) approximately 6000 nuclides can be located (approximately 3400 with β^- and approximately 2600 with β^+ decay). For the most part these nuclei are "exotic" – proton or neutron-rich nuclides. The real value of the exotic nuclei studies is very humble (if any) and the existing scientific interest in this area seems unjustified.

The proton stress predetermines the numbers of neutrons and the overall stability of the nucleus. As a measure for the proton stress (S_p), the ratio of the square of the proton numbers (Z^2) to the number of the neutrons in the neutron pairs (N_{2n}) could be used. The places between protons are equal to Z - 1, then $N_{2n} = N - (Z-1) = N - Z + 1$ = A - 2 Z + 1. Hence

$$S_p = \frac{Z^2}{A - 2Z + 1}$$

Fig. 17 shows the proton stress in stable nuclides of elements with proton numbers over 10. It is evident from Fig. 17 that proton stress can be used as a representative characteristic of the nuclei. The upper curve 1 represents the proton stress in even/even stable nuclides with minimum number of neutrons. The curve shows the ability of nuclei to resist to proton stress. The average value, equal to 170, is practically independent on the number of protons. The discrepancy is due to the "magic" numbers. The doubly "magic" number of Ca_{20}^{40} has the biggest resistance to proton stress $S_p = 400$, followed by $S_p = 262$ of Ni_{28}^{58} (the magic proton number). Surprisingly the next is Ru_{44}^{96} ($S_p = 214$), probably because the number of neutrons 52 is in a "magic" zone. The nuclide Cd_{48}^{106} endures a bigger proton stress than the nuclide Sn_{50}^{112}. After Z=50, the discrepancy becomes negligible, probably because the influence of the "magic" numbers on the intensity of nucleon oscillation decreases. Only Sm_{62}^{144} has higher proton stress (183) because of the neutron "magic" number 82.

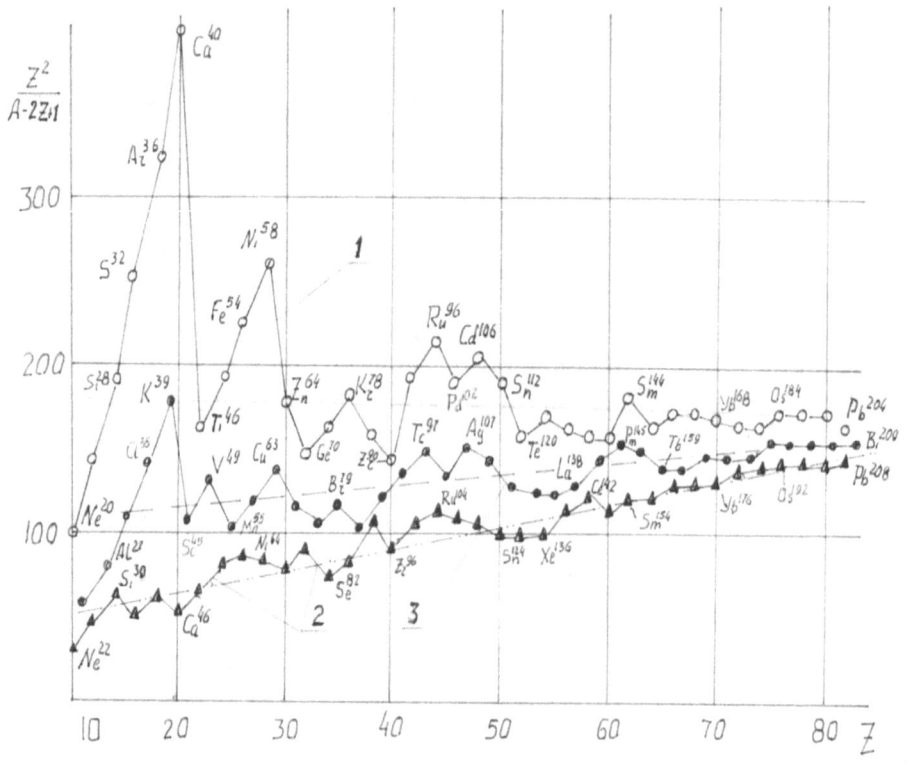

Fig. 17. Proton stress in stable nuclides
1. Even/even nuclides with minimum neutrons; 2. Even/even nuclides with maximum neutrons; 3.. Odd/even nuclides

Due to the nuclear asymmetry the odd/even nuclei can bear a smaller proton stress (the middle curve 3 on Fig. 17). The values are smaller and there is a trend for a slow increase from 110 to 156 (at

Bi_{83}^{209}). Fig. 17 shows that the proton stress in odd/even nuclides as a whole follows that of the even/even nuclei. The maximum (180) is the proton stress of K_{19}^{39} -"magic" neutron numbers and "near magic" number of protons. The relatively high values of proton stress of Ru_{44}^{96} and Cd_{48}^{106} in curve 1 are followed by high values of Tc_{43}^{97} (the most long lived isotope) and Ag_{47}^{107}. Evidently the nuclide Tc_{43}^{97} cannot stand such a high proton stress and is obliged to capture an electron. The next odd/even isotope Tc_{43}^{99} is in a zone of strong neutron stress and undergoes β^- decay. The nucleus of Ag_{47}^{107} is stable, because it has proton number near the "magic" one, 48. The reason for the Promethium instability is same as that of Technetium- the nucleus of Pm_{61}^{145} cannot stand high proton stress (S_p =144) and captures an electron. The next isotope, Pm_{61}^{147}, is in neutron stress zone and undergoes β^- decay.

The lowest curve 2 on Fig. 17 represents the proton stress in stable even/even nuclides with maximum number of neutrons. There is a well expressed increase of the proton stress from the average 50 (at Z= 10) to 147 of Pb_{82}^{208}. The discrepancy of values is considerably smaller. The smaller values have nuclides with "magic" nucleon numbers – oppositely to the values in the upper curve (proton stress at minimum neutrons). Only the proton stress of Ru_{44}^{104} (114) and Pd_{46}^{110} (113) are bigger than the average value of 95 for the region. A comparison with the location of Ru_{44}^{96} and Pd_{46}^{102} in the upper curve leads to a conclusion, that nuclei of Ruthenium and Palladium are more resistant to proton stress. Evidently the Technetium (which is in this region) cannot stand the elevated proton stress because of its nuclear asymmetry.

From fig. 17, it follows that some proton stress exists in all nuclei and it cannot be eliminated by neutron addition. In nuclides with maximum neutrons, the proton stress increases up and at Z = 92-95 it will reach the value of 170. Evidently when the proton stress of nuclides with maximum neutrons becomes equal to the proton stress in nuclides with minimum neutrons, the neutron shield becomes useless, and the nucleus becomes unstable.

4.2. BINDING ENERGY

The binding energy is one of the most expressive characteristics of the nucleus, but being a cumulative function makes it not so convenient for analysis of nuclei properties. Evidently any partial function would be much more convenient. It is the *binding energy per nucleon* (ε_N) that is in use in Nuclear Physics as a criterion for nuclei stability. It is well known, that the dependence of ε_N from the number of nucleons (A), has a maximum (ε_N= 8,79 MeV), which corresponds to A = 56 (the area of Iron). The slow decrease in ε_N at bigger A is explained as influence of electrostatic repulsion between protons. The cause for the ε_N dropping at light nuclides is not clear. If ε_N is used as a criterion for stability, *the Helium paradox* arises. The binding energy per nucleon for He_2^4 is only 7,074 MeV, less than 7,50 MeV for the heaviest unstable nuclides! But it is well known that the nucleus of He_2^4 is very tight and behaves as a particle (α -particle). It is proposed in the theory of α -decay, that in the heaviest, radioactive nuclei He_2^4 exists as a particle and moves there at the speed of light, i.e., a nucleus with less binding energy moves in the medium of more strongly bonded nucleons! *Evidently the binding energy per nucleon is not a correct characteristic.*

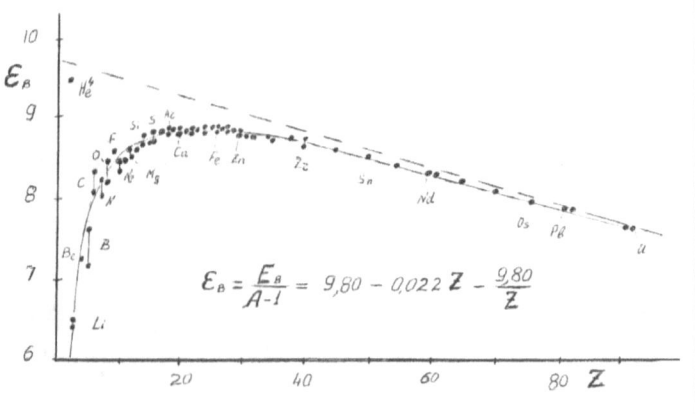

Fig. 18. Partial binding energy pro bond

As it can be seen from the schematics of arrangement of nucleons in the nucleus, the binding energy must not depend on the number of nucleons A, but on the number of ties between nucleons, which are equal to A − 1. On the Fig. 18, the *binding energy per bond* (ε_B) is shown as a function of proton number Z. It is evident from Fig. 18 that the curve of ε_B/Z dependence is similar to the well-known curve of ε_N/A dependence, but the value for He_2^4 is situated much higher than all others. This is the *solutions of the Helium paradox*. It is reasonable to accept the proton stress as the cause for ε_B decrease in heavier nuclides. But the cause for ε_B dropping in the lightest nuclides is quite incomprehensible. It cannot be a type of "intrinsic" property of all light nuclides because the lightest of them (He_2^4) has the biggest value of ε_B. Probably this anomaly is a consequence of the bond disturbances between nucleons in the middle of the nucleus. What will happen if two couples of **pn** are added consequently to a certain light nucleus? *A homologous series* of nuclides will be obtained. For instance, the homologous series on the basis of He_2^4 will be:

He_2^4 **pn:np**

Be_4^8 **pnpn:npnp**

C_6^{12} **pnpnpn:npnpnp**

O_8^{16} **pnpnpnpn:npnpnpnp**

and so on.

The regularities of changes in binding energy when two couples of **pn** are added are given in Tables 1 — 4. The basis for the calculations is the data for binding energy E_B taken from [1]. The results from the calculation for a homologous series of *even/even* nuclides are shown in Table 1. The energy brought by each **pn** pair is calculated as a difference between the binding energies (ΔE_B) of two neighboring nuclides. It is evident from Table 1 that in the frame of nuclear particularity ΔE_B is practically constant (35- 37 MeV). Please notice that this particularity does not coincide with assuming an influence of magic numbers. For instance, the maximum increase of ΔE_B is between Si_{14}^{28} and Mg_{12}^{24} with a number of nucleons far from the magic numbers 8 and 20. Only the difference between the binding energies of Be_4^8 and He_2^4 is considerably smaller – 28,20 MeV. The increase in energy related to one added nucleon ($\Delta E_B/4$) is also smaller (7,05 MeV) in comparison with the average 9 MeV for other nuclides.

Table 1. Binding energy in a series of even/even nuclides

N o	Nuclide	Binding energy, E_B, MeV	Bind. ener. difference, ΔE_B, MeV	$\frac{\Delta E_B}{4}$, MeV	$\varepsilon_B = \frac{E_B}{A-1}$ MeV	$\varepsilon'_B = \frac{E_B +}{A}$ MeV
1	He_2^4	28,30	–	–	9,43	
2	Be_4^8	56,50	28,20	7,05	8,07	9,32
3	C_6^{12}	92,16	35,66	8,91	8,38	9,17
4	O_8^{16}	127,62	35,46	8,87	8,51	9,11
5	Ne_{10}^{20}	160,65	33,03	8,25	8,45	8,91
6	Mg_{12}^{24}	198,26	37,61	9,40	8,62	9,00
7	Si_{14}^{28}	236,54	38,28	9,57	8,76	9,09
8	S_{16}^{32}	271,78	35,24	8,81	8,76	9,05
9	Ar_{18}^{36}	306,72	34,94	8,74	8,76	9,01
10	Ca_{20}^{40}	342,06	36,34	9,09	8,77	9,00
11	Ti_{22}^{44}	375,48	33,42	8,35	8,73	8,94
12	$Cr_{24,48}$	411,47	35,99	9,00	8,75	8,94
13	Fe_{26}^{52}	447,71	36,24	9,06	8,78	8,95
14	Ni_{28}^{56}	484,00	36,29	9,07	8,80	8,96

The binding energy per bond (ε_B = 8,07 MeV) is also smaller than the average value 8,5 – 8,7 MeV. The cause for this anomaly lies in the structure and the strength of He_2^4. The trend towards He_2^4 formation dominates and suppresses the formation of other ties between nucleons. This is also the cause for the surprisingly short half-life time (10^{-16} s) of Be_4^8. Practically, two nuclides of He_2^4 are formed instead of one Be_4^8. Thus one bond of approximately 8 – 9 MeV is lost. This loss is not a local particularity. It is present in all even/even nuclei, because a **pn:np** group exists in the center of each of them. Hence the problem turns out to be "inheritable". After Be_4^8, gradually a quantitative improvement of the characteristics begins, but the influence of the He_2^4 can be traced even until Ni_{28}^{56}. If we add the lacking energy 8,77 MeV to the binding energies of each nuclide after He_2^4, it would be possible to calculate the correlated binding energy per bond $\varepsilon_B' = (E_B + 8,77)/(A-1)$. It is evident from Table 1 that the values of ε_B' gradually decrease from 9,32 to 8,96 MeV due to the proton stress increase.

The cause for ε_B dropping at other light nuclides is also connected with central nucleon combination. Table 2 shows the binding energies in a homologous series of odd/odd nuclei on the basis of deuterium (H_1^2) :

H_1^2 **p:n**

Li_3^6 **pnp:nnp**

B_5^{10} **pnpnp:nnpnp**

and so on, according to Table 2.

Table 2. Binding energy in a series of odd/odd nuclides

No	Nuclide	Binding energy, E_B ,MeV	Difference, ΔE_B. MeV	$\dfrac{\Delta E_B}{4}$, MeV	$\varepsilon_B = \dfrac{E_B}{A-1}$ MeV	$\varepsilon_B' = \dfrac{E_B + 6}{A-}$ MeV
1	H_1^2	2,24	—	—	2,24	8,84
2	Li_3^6	31,99	29,65	7,41	6,38	7,72
3	B_5^{10}	64,75	32,76	8,19	7,19	7,92
4	N_7^{14}	104,66	39,91	9,98	8,05	8,56
5	F_9^{18}	137,37	32,71	8,19	8,08	8,47

6	Na_{11}^{22}	174,15	36,78	9,20	8,29	8,60
7	Al_{13}^{26}	211,90	37,75	9,43	8,48	8,73
8	P_{15}^{30}	250,61	38,71	9,67	8.64	8,87
9	Cl_{17}^{34}	285,57	34,98	8,75	8,65	8,85
10	K_{19}^{38}	320,64	35,07	8,75	8,67	8,84
11	Sc_{21}^{42}	354,68	34,04	8,50	8,65	8,81

The binding energy differences (ΔE) in this series are in the range of 33 – 39 MeV. Only between Li_3^6 and H_1^2 the ΔE is smaller (29,65 MeV). Evidently the cause lies in the relatively weak bond (2,24 MeV only) in H_1^2 (a **p:n** with parallel spins), and this weak bond lies in the center of all odd/odd nuclides. Should this energy be normally high (for instance 8,8 MeV) there would be no energy drop at light odd/odd nuclei. The values of ε_B' = (ΔE + 6,6) / (A-1) confirm this conclusion – all values are in the range of 7,72 – 8,87 MeV. The difference of 1 MeV is in the range of particularity of nuclides. Generally, however, the values for ε_B' are lower than those in even/even nuclides. This negative difference is comprehensible in view of instability of nuclides after N_7^{14}.

Table 3 shows the binding energies in a series of the odd/even nuclei. All nuclides in this series have Triton (H_1^3) as a base:

H_1^3 **p:n**

Li_3^7 **nnp:nnp**

B_5^{11} **npnnp:nnpnp**

and so on, according to Table 3.

Table 3. Binding energy in a series of odd/even nuclides

No	Nucli-de	Binding energy, E_B, MeV	Difference ΔE_B, MeV	$\dfrac{\Delta E_B}{4}$, MeV	$\varepsilon_B = \dfrac{E_B}{A-}$ MeV	$\varepsilon'_B = \dfrac{E_B+9}{A-1}$ MeV
1	H_1^3	8,48	—	—	4,24	8,75
2	Li_3^7	39,25	30,77	7,69	6,54	8,04
3	B_5^{11}	76,20	36,95	9,24	7,62	8,52
4	N_7^{15}	115,49	39,29	9,82	8,24	8,89
5	F_9^{19}	147,80	32,31	8,08	8,21	8,16
6	Na_{11}^{23}	186,57	37,77	9,44	8,48	8,43
7	Al_{13}^{27}	224,95	38,38	9,60	8,65	9,00
8	P_{15}^{31}	262,92	37,97	9,49	8,76	9,06
9	Cl_{17}^{35}	298,21	35,29	8,82	8,77	9,04

The binding energy of H_1^3 is 8,48 MeV and ε_B = 8,48/2 = 4,24 MeV. Consequently, the values of ε_B for Li_3^7 and B_5^{11} are bellow the average 8,60 MeV. By adding 9 MeV to the values of the binding energies, the values of ε'_B become comparable for all nuclides in the series.

Table 4 shows the series of even/odd nuclides, based on He_2^3:

He_2^3 **:np**

Be_4^7 **np:npnp**

C_6^{11} **npnp:npnpnp**

and so on, according to Table 4.

In this series the influence of the small binding energy in He_2^3 reflects also on the binding energy of the rest of the nuclides. By adding 10 MeV to the binding energies, the values of ε'_B become practically identical (in the range of approximately 1 MeV).

Table 4. Binding energy in a series of even/odd nuclides

No	Nuclide	Binding energy, E_B, MeV	Difference ΔE_B, MeV	$\dfrac{\Delta E_B}{4}$ MeV	$\varepsilon_B = \dfrac{E_B}{A-}$ MeV	$\varepsilon'_B = \dfrac{E_B}{A}$ MeV
1	He_2^3	7,72	–	–	3,84	8,85
2	Be_4^7	37,60	29,88	7,47	6,27	7,93
3	C_6^{11}	73,44	35,84	8,96	7,34	8,34
4	O_8^{15}	111,95	34,51	8,63	8,00	8,71
5	Ne_{10}^{19}	143,78	31,83	7,95	7,99	8,56
6	Mg_{12}^{23}	181,73	37,95	9,49	8,26	8,72
7	Si_{14}^{27}	219,36	37,63	9,41	8,44	8,82
8	S_{16}^{31}	265,70	37,34	9,36	8,56	8,89
9	Ar_{18}^{35}	291,47	34,77	8,69	8,57	8,87
10	Ca_{20}^{39}	326,42	34,95	8,75	8,59	8,85

The calculation above is sufficient to prove the hypothesis that the low binding energy in light nuclei is inheritable – it follows from the weak bonds in the middle of the nuclei.

4.3. RADIOACTIVITY

Radioactivity is a process of particle release from nuclei. As a result the initial (or *mother*) nucleus transforms into another (*daughter*) nucleus with different properties. Very often the radioactivity is accompanied by electromagnetic (γ) emission.

The Natural radioactivity consists of releasing particles (electrons and He_2^4-nuclei) and nuclei fission of the existing in Nature isotopes. *The artificial radioactivity* concerns decay of artificially created nuclei, which, as more unstable, can release in addition protons, neutrons and groups of nucleons. In addition to releasing of particles, at certain conditions the nucleus can be simply destroyed. The conditional border between the radioactivity and nucleus destruction is the half-life time. In radioactivity the half-life ranges from milliseconds (ms) to billion years. In nucleus destruction, the half-life is in the range of $10^{-12} - 10^{-14} s$. It is thought that the nucleus destruction passes through a state of *compound nucleus*, formed by obtaining a considerable portion of energy from outside. But with respect to the energy both processes are equivalent. The difference in half-life is due to the difference in energy available within the nucleus. As a rule, the more energy available, the shorter the half-life.

It is well known, that radioactivity is a *random process* and in the Theory of Probability [15] it is cited as an example of *Homogeneous stochastic process with independent increment*. Moreover, the radioactive decay possesses *stationarity, ordinariness, absence of aftereffects* and hence relates to *a Poisson processes*, where the probability $P_{n(t)}$ for n decays to occur in time t is:

$$P_{n(t)} = \frac{(\lambda.t)^n}{n!} \exp(-\lambda.t)$$

where λ is the constant of decay.

Besides the rate of the process which can be expressed by this or by similar equations, it is very important to obtain more information about the mechanism of the decay. Although the mechanism of each kind of radioactivity is unique, there are some common characteristic features. Let us compare the modes of decay and half-life time of nuclides in the following homologous series:

$Se_{34}^{86}(\beta^-,16s); Kr_{36}^{90}(\beta^-,\gamma,33s); Sr_{38}^{94}(\beta^-,\gamma,76s); Zr_{40}^{98}(\beta^-,31s); Mo_{42}^{102}(\beta^-,62s);$

$Ru_{44}^{106}(\beta^-,369days); Pd_{46}^{110} - Ba_{56}^{130}(6 stable.nuclides); Se_{58}^{134}(\varepsilon.\gamma,72h); Nd_{60}^{138}(\varepsilon,\gamma,5h)$

$Sm_{62}^{142}(\varepsilon,\beta^+,\gamma,73\min); Gd_{64}^{146}(\varepsilon,\beta^+,\gamma,50days); Dy_{66}^{150}(\beta^+,\varepsilon,\gamma,\alpha,7,2\min.)$

$Er_{68}^{154}(\alpha,5\min); Y_{70}^{158}(\alpha,1s)$.

It is clear enough that a gradual increase in the number of protons and neutrons (by two **pn** couples) leads to gradual change in the spectrum of radioactivity from β^- decay, through the stable state , β^+ decay and electron capture(ε), to α decay. Formally the increase of **pn** couples leads only to the change in neutrons/protons ratio from $52/34 = 1,53$ to $88/70 = 1,26$ which causes the changes in the proton stress. Hence the entire spectrum of radioactivity is a consequence of energy imbalances.

The energy balance in radioactive decay consists of tree items:

1. Energy necessary for *the bonds breaking* (E_n), where n is the number of portions (quanta) of energy.

2. Energy available within the nucleus expressed as average amount of energy related to one nucleon, i.e. *the mathematical expectation* (E_0).

3. Energy which will *be gained* in the process ΔE (equal to the difference of binding energy of the daughter's and mother's nuclei).

Each nucleus having one or more weak ties is predisposed to decay. The moment of decay is a question of storing up of enough energy on a weak bond. The role of energy carrier is played by the oscillation. In my book "The Secret of Creation" [16] it was shown that particles which obey Bose statistics (as the energy quanta) follow an Exponential distribution:

$$P_{(En)} = \frac{1}{1 + E_0} \cdot \exp(-\frac{E_n}{E_0})$$

Where $P_{(En)}$ is the probability for n energy quanta to fall on a definite bond. This probability is proportional to the constant of decay (λ), (or inversely proportional to the half-life). The energy ΔE does not take a direct part in the equation above, but it is a "guarantor" for the irreversibility of the process.

The process of energy redistribution shows that there is a certain probability for any quanta of energy falling on a definite bond, which will lead to an act of radioactivity. Hence *the process of energy redistribution in nucleus is equivalent of the quantum effect "tunneling through energetic barrier"*.

It is not difficult to understand, that all experimentally settled regularity for half-life dependence on energy (for α and β radioactivity), such as the rule of Geiger-Natal, can be deducted from the equation above. Of course, the statistical equations are good for practical calculation, but they do not concern the mechanism of decay, which is different in different kinds of radioactivity.

4.3.1 Alpha decay

Alpha decay is a process of He_2^4 nuclides (or α- particles) releasing. Modern Physics explains alpha decay as He_2^4 nuclei formation in the volume of the decaying spherical nucleus, going to and fro (10^{20} times in sec) in that spherical volume, and getting out through the surface by tunneling effect. Perhaps it looks very sophisticated, but it is not true. In the middle of a colossal density (tightly tied nucleons) a motion of such big particle with the speed of light is absurd. It would need a colossal driving force and would destroy all links in the nucleus (see nuclear reactions with α - particles). Moreover, if formation of one α - particle is possible (from energy point of view), why the formation of more α - particles should be impossible?

It is evident from the homologous series above, that α - decay occurs in nuclides having an increased proton stress. Hence the proton repulsion must play the basic role in tearing off the He_2^4 nucleus. The He_2^4 nucleus can be formed at both ends of nuclei by charge and spin exchanges between protons and neutrons. For instance, if the left branch of a nucleus ends as **pnpnnpn...** it is only necessary to exchange a charge between the second proton and second neutron in order to form α - particles. After tearing off an α - particle, at the end of the left branch of the daughter's nucleus charge and spin exchanges occurs and **npn...**group becomes **pnn...** If a nucleus ends on **pnpnpn...** more energy is necessary for the charge exchange between the second proton and second neutron because of the repulsion from the third proton. This is the cause for light and middle mass nuclei resistance to α - decay. When the branch ends on a **pnnpnn...** group, α - particle formation becomes more difficult because it needs more exchanges of spins and charges. This is why in neutron- rich nuclides α - decay gradually becomes replaced by β^- - decay.

Formally the α - particle tearing off becomes possible only in nuclei whose binding energy per bond is less than 28,3/4 = 7,1 MeV. But even in the heaviest nuclei the binding energy per bond is higher (7,6 MeV for U_{92}^{238}). As it was said above, the oscillation of the nucleons inside the nucleus causes redistribution of the binding energy. Hence the tearing off an α - particle occurs when necessary quantity of energy comes on the bond of the first **nn** pair from the end of a branch (let us mark it by an accent, **pnpn'npn...**). The probability of these events to occur can be estimated by the equation above if the necessary energy (E_n) and the mathematical expectation (Eo) are known.

The kinetic energy of an α - particle is a sum of a part of binding energy spent on the tearing, the energy obtained from the electrostatic repulsion and the kinetic energy of wave motion in the nucleus at the ends of the branch. This explains why the kinetic energy of α-particles of even/even nuclei (especially those with "magic" numbers of nuclides) is smaller than the kinetic energy of the rest of the nuclei.

4.3.2. Beta decay

Usually the positive and negative beta decay as well as the electron capture are considered as a unique process of weak interaction. Formally it may be correct, but actually there is a considerable difference between them. The neutron spontaneously decays and releases particles and energy, but the proton does not decay at all. The proton only helps *the energy decay* (decay of a symmetric wave in two asymmetric waves) and this essential difference in physics of the processes must not be neglected. Each process needs a separate analysis.

4.3.2.1. β^- - decay

The reaction $C^{14} \rightarrow N^{14} + e^- + \widetilde{\nu}$ can be accepted as typical for β^- decay.

A nucleus of C_6^{14} decays by emitting an electron (e^-) and an antineutrino ($\widetilde{\nu}$). The half-life is 5730 y, and the decay is not accompanied by γ - emission.

The question "why the nucleus undergoes β^- decay" is incorrect, because the neutrons in free states are unstable. The reasonable question must be: "why the neutrons in nuclei do not decay"? Does the weak interaction in nuclei become stronger? Evidently the condition within nuclei can promote, delay or exclude the neutron decay. We will try to present a schematic illustration of the cause for β^- decay without involving the "boson fields".

As it was pointed out earlier, the neutron consists of a proton and a negatively charged wave located in a proton potential field. The bond between the two opposite waves is purely electrostatic. As the asymmetry of the negative wave does not match the radius of the proton, the trend towards wave straightening leads to neutron decay. The basic bond (strong interaction) in a nucleus is between the waves of the two protons as it is shown in Fig. 4 and Fig. 5. Generally, in nuclei with suitable arrangement of nucleons, a supplementary bond is established between the fields of a negative wave of the neutron and the neighboring proton. Hence the negative wave becomes connected by two protons and so the "weak interaction" in nuclei becomes twice as strong as in a free neutron ($f = e^-.2e^+ = 2e^-.e^+$). This is the cause for the *neutron stability* in nuclei.

The cause for *the neutron decay delay* in nuclei is the shortage of the electrostatic bonds of neutrons which are out of **pn** pairs. One proton is unable to bind tightly two negative waves (of two neutrons, as for instance in H_1^3). In favorable energy condition one of the negative waves being loosely tied straightens and separates from the nucleus by forming electron and neutrino.

The cause for the *neutron decay acceleration* (short half-life) in nuclei is the repulsion between the negative waves of neighboring neutrons (the neutron stress). This repulsion helps in ties tearing (between the negative wave and positive electrostatic field of the proton) and accelerates the neutron decay.

As the neutron formation occurs with energy absorption, the binding energy difference between the "daughter's" and "mother's" nuclei in β^- decay can be negative. For instance the nuclide of Tritium has a binding energy of 8,48 MeV. The "daughters" nuclide He_2^3 has a binding energy of 7,72 MeV. The difference is negative – 0,76 MeV. In the decay of a free neutron the total mass release is 1,29 MeV, including 0,51 MeV mass of the electron. Hence the rest of the energy 1,29 – 0,51 = 0,78 MeV is just enough to compensate the shortage of energy above. The negative energy in β^- decay predetermines a long half-life, for instance: $C^{14} \rightarrow N^{14} - 0,63 MeV, (5730 \quad y); \quad Cu^{63} \rightarrow Ni^{63} - 0,71 MeV, (92y);$ $Zr^{93} \rightarrow Nb^{93} - 0,72 MeV, (10^6 y).$

Generally the increase of positive energy leads to shortened half-life and β^- decay is accompanied by γ emission. The changes of energy are bigger and respectively the half-lives are shorter in β^- decay of light mass nuclides, for instance:

$Be^{11} \rightarrow B^{11} + 10,73 MeV, (13,6s);$

$N^{18} \rightarrow O^{18} + 13,27 MeV, (0,63s).$

4.3.2.2. β^+ - decay

Very often the basic β^+ - decay is referred to as "proton decay"

$p \rightarrow n + e^+ + v$

This interpretation leads to the wrong conclusion that each proton can release a positron (e^+) and a neutrino (v). If this were correct, why the protons in Hydrogen do not decay? No, the protons cannot decay (even after 10^{31} years- as the "standard" theory predicts). Evidently, something very important is omitted in the reaction above.

To be emitted, the positron and neutrino have to be created from something. A portion of energy must be supplied to the proton, most probably as γ - quanta. The symmetric electromagnetic wave cannot be kept by the proton, but the electrostatic field can split the γ - wave in two asymmetric waves. First, the negatively charged one interacts with the positive field of the proton, forming a neutron. The second, the positively charged one, quickly separates from the proton, forming a positron and a neutrino. Hence, the correct reaction of β^+ - decay should be

$\gamma + p \rightarrow n + e^+ + v$

The γ - energy must be at least twice as much as the difference between the masses of the neutron and proton, i.e., 2. 1,29 = 2,58 MeV. One of the lowest energy values 2,77 MeV is in β^+ - decay of C_6^{11} (260 min half-life). One of the highest energy values 16,1 MeV is in β^+ - decay of Sc_{21}^{40} (0,18 s half-life).

The cause for β^+ - decay in the nuclei is the proton stress. The repulsion between electrostatic fields of protons reduces the ties between nucleons, which is equivalent to concentration of energy. When this energy reaches the minimal necessary value of 2,58 MeV, it can be captured by the proton electric field and transforms in two oppositely charged asymmetric waves.

4.3.2.3. Electron capture (ε)

Usually the process of electron capture is represented as:

$e^- + p \rightarrow n + v$

And if this neutron decays $n \rightarrow p + e^- + \widetilde{\nu}$, the common reaction will be:

$$e^- + p \rightarrow p + e^- + \nu + \widetilde{\nu}$$

Where does the neutrino and antineutrino come from? It will be more correct if the role of energy in each of reactions is specified. These processes include the reaction of wave (energy) splitting:

$$\gamma + p \rightarrow p + \nu + \widetilde{\nu}$$

The electron capture is a process rival to β^+- decay, and it usually wins the competition at lower proton stress (low energy) which corresponds to a long half-life. It is because the electron capture needs less energy than β^+- decay. As the electron it self presents a portion of 051 MeV energy, the electron capture needs only 0,78 MeV (as a minimum) energy. This energy must be available at energy redistribution within the nucleus and this condition is easy to fulfill and the eventual excess of available energy will be released during the decay. The lowest energy (0,86 MeV) is released in the electron capture of Te^{123} (10^{12} y half-life).

4.3.2.4. Mixed ($\beta^-, \beta^+, \varepsilon$) decays

It is well known, that some nuclei undergo mixed ($\beta^-, \beta^+, \varepsilon$) decay, but the *cause of such mixture of contradictory processes is still a puzzle*. The process looks really nonsensical- the protons and neutrons in a nucleus have opposite trends: the neutrons tend to release electrons and to become protons, and the protons tend to accept electrons and to become neutrons! If such a trend really exists in a nucleus, the easiest way for it to happen is through charge exchange – a process common to all nuclear reactions. The solution of the puzzle is possible only in the frame of the proposed structure of the nucleus.

Only odd/odd nuclei undergo a mixed decay. They are just before the stable odd/even nuclei, or between the two stable odd/even nuclei (when two of them exist), for instance, $Cl_{17}^{36}, K_{19}^{40}, V_{23}^{50}$. The cause is the double asymmetry in odd/odd nuclei – one extra proton and one extra neutron. From the point of view of the number of nucleons, each odd/odd nucleus consists practically of branches which belong to two different even/even nuclei. For instance, the Z+1 proton branch of Cl_{17}^{36} is equal to the branches of Ar_{18}^{38}, and the Z – 1 proton branch is equal to the branches of S_{16}^{34} . Thus odd/odd nuclei have two branches with opposite type of stress. The proton stress prevails in the branch with extra proton, and it undergoes an electron capture or a β^+ decay. In the other branch the neutron stress prevails and it undergoes a β^- decay.

4.3.3. Proton decay

The proton decay consists of tearing off a proton from a nucleus. This process occurs only in the artificially created proton-rich nuclei, whose structure comprises several pairs of protons, and one of the pairs must be at the end of a branch. The force of electrostatic repulsion is the main cause for the proton tearing off, although the contribution of the destructive force of asymmetry cannot be neglected. Generally, proton decay is a rival process to β^+ decay. The speed of the processes decides the outcome of the competition.

The proton decay in light even/odd nuclei occurs when the excess of protons over neutrons is equal to three ($Z - N = 3$). In this case 2 pairs of protons exist in the nucleus, one in the center and the other at the end of a chain. For instance the structure of the nuclide O_8^{13} is **ppnpnp:nppnpnp.** The first proton from the left breaks off. The left proton in the couple **pp** in the right branch undergoes a β^+ decay and the final result (the daughter nucleus) is C_6^{12} **pnpnpn:npnpnp.**

The role of asymmetry can be seen in the following example. The even/odd nuclide Mg_{12}^{21} has two **pp** couples – one in the center and one at the end of the other branch (like in the nuclide O_8^{13}) and undergoes a β^+, p decay with 0,12 s half-life. The even/even nuclide Mg_{12}^{20} has 3 **pp** pairs – one in the center and two at both ends. Despite the stronger proton stress (stronger repulsion), the absence of asymmetry-related destructive force only leads to β^+ decay with a longer half-life (0,60 s). The even/even nuclide Mg_{12}^{22} has only one **pp** in the center and decays in β^+, γ with 4 s half-life. Hence, in the absence of asymmetry, a pair of protons (**pp**), despite the electrostatic repulsion, can exist in a nucleus for several sec.

With increasing number Z, it becomes possible for the proton decay to occur at smaller proton excess, and even at small neutron excess (in the heaviest nuclei).

Two proton decay is a release of 2 protons simultaneously or in a sequence of times shorter than the half-life. Probably, the process of a simultaneous release of two protons can take place in even/even nuclei, (in the absence of asymmetry) and with a proton excess bigger than that in the case of one proton decay – i.e., when there are a minimum of 5 **pp** couples, two of which are at the ends of chains.

4.3.4. Neutron decay

The cause of neutron release from nuclei is the neutron stress. It occurs in the neutron-rich nuclei. In exotic nuclei, (with a considerable excess of neutrons), the neutrons behind the last protons in the branches part easily from the nucleus. For instance, Be_4^{12} undergoes a β^-, n decay with 11,4 ms half-life. The structure of Be_4^{12} is **npnnpn:npnnpn**. One of the final neutrons drops off, the other undergoes a β^- decay simultaneously with a charge exchange between the neighboring **np** pair.

With smaller neutron excess, the process of neutron release occurs via charge exchange. For instance, C_6^{16} undergoes a β^-, n decay, but its structure is **pnnpnnpn:npnnpnnp.** Under the neutron stress, a charge exchange occurs between the first **pn** pair and between proton number 5 and neutron number 7 from the left. The structure becomes: **npnpnnpn:npnpnnnp.** The first neutron drops off. The middle neutron in the group at the other end **...pnnnp** undergoes β^- decay. The resulting nucleus is N_7^{15}.

Two neutron decay is probably possible for a bigger neutron excess in even/even nuclei. But in light nuclei the number of neutrons cannot exceed 2Z + 2. Hence the simultaneous release of 2 neutrons could only be a decay of "exotic" nuclei near to neutron drip line.

4.3.5. Spontaneous fission of nuclei

The nuclei of the heaviest elements ($Z \geq 90$) undergo spontaneous fission. The Modern Physics postulates that spontaneous fission occurs via tunneling effect through an energetic barrier. As it was shown above, the physical equivalent of the tunneling effect is the energy redistribution in nucleus.

The spontaneous fission is a type of radioactivity and thus it obeys the basic law of radioactive decay. The cause of the spontaneous fission is similar to the cause for α - decay. Energy redistribution (during nuclei oscillation) creates the possibility that a certain weak bond can get more energy, enough to tear the nucleus apart at a certain moment. The probability for this event to occur increases with time. The half-life time is shorter when there is larger available energy (bigger mathematical expectation). It is well known [3], that the speed of spontaneous fission of some heavy isomeric nuclides is approximately 10^{80} times higher than the speed of spontaneous fission of the same nuclides in ground state. Such speed can be achieved if the energy in the basic state is in the KeV range and in the isomeric state - in the range of 200 KeV – very reasonable values.

From physical point of view the spontaneous fission is the result of the extension of the nucleus due to increased proton stress. The proton stress maximum is in the middle of the nucleus, where the bond is weaker, and this predetermines that the fragments would have the same size. Besides, the proton stress causes charge exchanges. For instance, the usual structure in the middle of the heaviest nuclei is: **...npn:npn...** The proton stress increase provokes charge exchange: **...pnn:npn...** which leads to formation of three neutrons side by side and to destruction of the nucleus. The neutrons act as a screen and reduce the proton stress, but they also reduce the strength of the ties. As a result the half-life of proton-rich nuclei increases with increasing number of neutrons, reaching a maximum at a certain Z^2 / N ratio and decreases with further increase in the number of neutrons. Generally, symmetric (even/even) nuclei have a shorter half-life, because more energy is available in the center. In asymmetric nuclei part of the energy is stored in the "orbits" (de Broglie waves) and at the middle of the branches as energy of oscillation, rotation, etc. In such cases, the nature of the induced (non spontaneous) fission is different and will be discussed further.

4.3.6. Gamma emission and nuclear isomerism

Gamma emission is a way of leveling the energy imbalance in radioactivity and nuclear reactions. Gamma emission occurs mainly when there are changes of binding energy. As it was established in the previous Chapters, the ties of the strong interaction are due to the interference of a part of the matter waves. The increase in binding energy, with shortening of the polarized chains, leads to the formation of gamma rays – electromagnetic waves, which consist of polarized chains. Another source of gamma emission in the nucleus is changes in wave length of "orbits". A decrease of the energy of the de Broglie wave ("orbit") can occur via formation and release of electromagnetic wave. Of course, both ways are mutually dependent. Probably the usual way of the de Broglie wave energy decrease happens via energy increase of nucleon oscillation and subsequent energy release as gamma rays. That is why, as a rule, each act of gamma emission is connected with nuclear spin changes. But this dependence is not very strict and the deviations are the basis for *nuclear isomerism*.

Usually the nuclei formed as products of nuclear reactions are in ground state with certain minimum energy of nucleon oscillation and corresponding minimum spin energy. But some times the energy equilibration is not fast enough and for a certain period of time the nucleus remains in exited or isomeric state – able to release gamma energy. The one-level isomeric state is typical for odd/even and even/odd nuclides. Some of the odd/odd nuclides, which undergo the mixed β^+, β^- decay, have two isomeric levels. The release of the gamma energy can occur sequentially from the top to the middle and to the ground level (for inst. Br_{35}^{08}) or simultaneously from the top to the middle and to the ground level, but with different half-life (for inst. Ir_{77}^{192}). The existence of two isomeric levels follows naturally from the structure of these nuclides. The explanation of mixed decay comes from the fact that each odd/odd nuclide consists of branches of two different nuclei. Hence each of these branches can have different properties (one undergoes β^+ and the other - β^- decay) and different isomeric levels.

4.4. NUCLEAR REACTIONS

Nuclear reactions cover various transformations of nuclei caused by collision with particles (p,n,d,α etc), γ - quanta, or by collision between nuclei. If a particle a collides with a nucleus A and as a result a separation of particles b and c occurs, and a new nucleus B is formed, the reaction can be shortly written down as $A(a,bc)B$. It is experimentally established that nuclear reactions are not strictly regulated. During a collision of A and a, different products can be obtained, and the nucleus B can be obtained in a collision of other nuclei and other particles. The usual cross-section of nuclear reactions is on the order of $10^{-27} - 10^{-21} cm^2$ and the usual energy of the particles is over 10 MeV (necessary for overcoming the electrostatic repulsion). Usually the necessity for understanding the structure of the nucleus is given as the main objective for studying the nuclear reactions. The basic nuclear reactions were well studied long ago but the nuclear structure is still not understood. Thus it is more reasonable to use the nuclear structure to explain nuclear reactions.

A general view of nuclear reactions shows that it is well advised to underscore the difference between reactions caused by an external force (*mechanical*) and reactions predisposed by *structural* peculiarity. Although in both reactions the energy plays a fundamental role, this role is especially crucial in the "mechanical" reaction. The structure of the nucleus and its elements (nucleons and groups of nucleons) plays a secondary role.

The "mechanical" nuclear reaction. As the binding energy between each nucleon pair is responsible for the stability of the nucleus as a whole, the nuclear reaction must be very sensitive to the energy of the colliding particles:

At low energy – *elastic scattering*. The energy is insufficient to tear a nucleon from the nucleus.

At middle energy – *the direct process*. The particle has enough energy necessary for tearing off the edmost proton, or the edmost group **pn...** -as a deuteron.

At higher energy – *spallation*. The more energy, the more damage to the structure of the nucleus. The number of particles in the product should be roughly proportional to the ratio of the amount of energy to the binding energy per bond.

The "structural" nuclear reaction. Structural nuclear reactions can be divided in two groups: 1. *Nuclear fission* – a process of heavy nucleus decay into two middle mass nuclei with release of neutrons and energy. 2. *Nuclear fusion* – process of two light nuclei conversion into one nucleus, accompanied by energy release.

4.4.1. Nuclear fission

Nuclear fission, as a leading process in nuclear physics, needs a more detailed consideration. It is evident that this process is induced by the neutron capture and depends on the neutron energy. The process could be treated as a fission by thermal neutrons (which have energy below 1 eV) and as a fission by fast neutrons (with energy above 1 MeV). This is all that is evident and it is the theory that has to explain the rest. The explanation of the essence of the process, according to [10], is as follows: A nucleus (for instance U_{92}^{235}) absorbs a thermal neutron, producing a compound nucleus (respectively U_{92}^{236}) in a highly excited state which undergoes fission – splitting into two fragments .These fragments rapidly emit *prompt neutrons*. The model of nuclear fission, developed by N. Bohor and J. Wheelr [10], is based on the liquid drop model and assumes that the thermal neutron falls into a potential well associated with the strong nuclear forces that act in the nuclear interior. Its potential energy (equal to the binding energy of a nucleon) transforms into excitation energy. Under the influence of this energy the nucleus, behaving like an energetically-oscillating charged liquid drop, will sooner or later develop a short "neck" and will begin to separate into two charged globs. The separation can only occur when a potential barrier is tunneled. This picture looks very sophisticated, but is too far from the truth. First of all, if the nuclear fission depends on the excitation energy, the speed of the process must increase with increasing the energy of the thermal neutrons. But it is well known that the cross-section of the fission with thermal neutrons decreases with increasing energy of the neutrons. Does the energy play a negative role! In addition, not all thermal neutrons are able to cause fission. The absorption of many neutrons ends by the formation of a U_{92}^{236} nucleus in ground state. Secondly, the slow neutrons can cause fission only in heavy mass ($Z \geq 90$) nuclei, which have neutron asymmetry – an extra neutron as, for instance, $U_{92}^{233}, U_{92}^{235}, Pu_{94}^{230}, Pa_{91}^{232}, Ne_{93}^{236}, Ne_{93}^{238}, Am_{95}^{242}$. And finally, the question about the mass asymmetry of the fragments and the number of the prompt neutrons in the fission are not at all discussed. Evidently an absence of adequate nuclear model leads only to inadequate solutions.

Both fission processes are natural extension of the spontaneous fission, considered above. The role of the neutrons is to cause some changes in the structure of the nucleus (increasing the asymmetry and formation of a weak bond) and respectively to bring in the nucleus supplementary energy (and thus to increase the mathematical expectation). When the energy of the neutrons is above 8 –10 MeV, the fission becomes a "mechanical" reaction. But despite the analogy, there is a quantitative difference.

Nuclear fission by thermal neutrons is a process of nucleus destruction only by *the destructive force of two neutrons asymmetry.* The nucleus cannot stand the violent nucleon oscillation and separates into two fragments. That is why only nuclei having one extra neutron (even/odd and odd/odd nuclei) undergo fission by thermal neutrons. For a nuclear fission to occur, it is necessary that a thermal neutron from outside to get into the branch having the extra neutron. But not all neutrons getting into that branch can cause fission. The nucleus resists the destructive force by structural changes - by "sending" extra nucleons in "orbit", or by sending a neutron in the shorter branch through spin and charge exchanges. The probability of the nuclear fission to occur varies along the branch. Except that the probability for penetration of the thermal neutron is not the same for all places along the branch. The total probability predetermines the ratio between the fragment masses. The thermal neutron penetrates easier in the places with bigger amplitude of nucleon oscillation, where the ties are weaker and, accordingly, the probability for such neutron to cause fission is higher.

Let us analyze the fission of U_{92}^{235}. The nucleus of U_{92}^{235} consists of two unequal branches. The shorter is equivalent to a branch of U_{92}^{234}, and the larger – to a branch of U_{92}^{236}. The difference between branches (or the asymmetry) is equal to one neutron. This asymmetry is sufficient to predispose the nucleus U_{92}^{235} to fission. The larger branch of the nucleus U_{92}^{235} is **16(pnpnn)3(pnnpnpnn)4(pnn)pn:pn...** The middle of this branch can be represented in details as follows:
...pnpnnpnpnnpnpnn[pnnpnpnnpnnpnpnnpnnpnpnn]pnnpnnpnn pnnpn:np...

27	29	31	32	33	35	36	37	38	39	40	41	42	43	44	45	46	47
65			88			89			95			107			118			
10^{-6}			0,02			0,65			1			0,02			0,002			

The digits in the first row are the enumeration of protons counted from the left end of the branch. The possible masses of the smaller fission fragment are given in the second row. The third row shows the relative total probability of nuclear fission.

It is evident that up to proton number 32 the structure of U_{92}^{235} is **pnpnnpnpnn...** This is the ordinary, more stable arrangement of nucleons, difficult for the thermal neutrons to penetrate. The probability of nuclear fission to occur after an accidental neutron penetration is also very small. As a result, the formation of fragments with mass less than 80 is relatively rare. After proton number 42, the structure is **...pnnpnn...,** a structure saturated with neutrons, not predisposed for accepting more neutrons, but practically each neutron penetration there leads to nuclear fission. The smaller probability of neutron penetration predetermines the relatively smaller probability for the resulting fragments to have mass in the range of 108 - 118. Between protons 32 and 42, (the range in the square brackets), is: **...pnnpnpnn...** This a transitive structure, with irregular neutron stress along the branch. It leads to formation of large gaps between the nucleons – a condition favorable for easier thermal neutron penetration. But by acceptance of one more neutron, the branch becomes alike with a branch of U_{92}^{238}. The excess of two neutrons causes the arising of a strong destructive force of nucleon oscillation which overcomes the forces of the strong interaction, and tears the nucleus into two fragments.

In the shorter branch of U_{92}^{235} (which is like a branch of U_{92}^{234}), the transitive zone is shorter and, as a result, the probability of getting a neutron there is approximately 6 time smaller than the respective probability of it getting into the larger branch. After capturing a neutron the shorter branch turns into a longer branch (like a branch in U_{92}^{236}) and thus the asymmetric nucleus of U_{92}^{235} becomes a symmetric nucleus of U_{92}^{236}, which does not undergo the fission by thermal neutrons.

Neutron releasing at fission. It is well known that at nuclear fission two sort of neutrons are released: a) Prompt neutrons (from 1 to 7) owing to which the existence of self- sustaining chain reactions is possible; b) Delayed neutrons – a small part of all neutrons, which appear latter and are of significance for the controllability in the operation of nuclear reaction. Prompt neutron separation occurs in the moment of the nuclear fission while the delayed neutrinos releasing is e result of radioactive decay of fission fragments.

From the point of view of the existing models the neutron release in nuclear fission is a real puzzle: At the division of liquid drop or liquid core by a neck formation, it is logical to expect a proton separation. The protons in the neck being loosely tied have to be repulsed from the positive charges of new formed drops! But instead of charged protons, the electro-neutral nucleons are released and it is not because of the neutron excess. It is well known that the balance of neutrons in fragments occurs by a sequence of β^- - decays.

The number of prompt neutrons depends on two conditions: 1. The place of the thermal neutron penetration. 2. The phases of nucleon and charge oscillations at the moment of fission.

When the nuclear fission follows immediately the act of neutron absorption (the usual way of fission) the number of the prompt neutrons is 2 or 3 – depending on the place of the neutron penetration in the longer nuclear branch. As all stable nuclei, the long living isotopes obtained as products of fission, must have protons at the ends of the branches. The neutrons which have been at the place of the nuclear tearing have to be released immediately. So, if the thermal neutron enters in a group … **pnnp…** three neutrons should be released. If the group is **…pnp…** the released neutrons should be two. In U_{92}^{235} fission the average number of released neutrons is 2,44 . This testifies for the fact that the probability for a thermal neutron getting into a **pnp** group is twice as high as the probability of its getting into a **pnnp** group. At U_{92}^{233} fission the ratio of probabilities is practically the same (average 2,3 prompt neutrons), but at Pu_{94}^{239} fission (2,09 prompt neutrons), the probability for a neutron getting into a **pnnp** group is very small. The role of the higher neutron stress in a **pnnp** group becomes evident.

But some time the fission does not occur immediately the thermal neutron absorption. The nucleus prove to be able to resist the fission for several acts of nucleons oscillation. Meanwhile, the nucleon oscillation is usually accompanied by a strong charge oscillation (a process of temporary negative neutron wave passing to the neighboring nucleons). The number of prompt neutrons (in the range 1 – 7) depends mainly on the phase of the charges oscillation in the moment of the fission. For instance, if the thermal neutron penetrates in the middle of the group ... *pnnpn[n]hphhp* ...(in brackets is shown the thermal neutron), because of the charge oscillation (direction shown by arrows) the group temporarily becomes ... *npnnp[n]pnnpn* ... If in that moment the nuclear fission occur, only one prompt neutron will be release (this in the brackets). But at reverse charges oscillation the group ... *phhphhpn[n]npnnpnnp* ... becomes ... *pnpnpnnnnnnnnpnpnp* ... If the nuclear fission occurs in that moment all 7 neutron in the center must be released, because the fragments ought to end by proton.

The charge oscillation can continue in the fragment and is responsible for the delayed neutron releasing. For that to occur it is necessary the newly formed branch of the fragment to end by a group *pnnpnnp* ... Because of the charge oscillation the group becomes *npnpnnp* ... and after separation of the endmost neutron the fragment become in more stable configuration.

Usually the speed of a nuclear reaction is measured by its cross-section. When the neutron energy increases up to 0,5 MeV, the cross-section of U_{92}^{135} fission decreases from approximately 2.10^{-24} cm^2 to approximately $1,3.10^{-24}$ cm^2 [3], because of increasing probability of neutron getting into other regions, where the thermal neutrons are unable to cause fission of the nucleus. In the range of approximately 0,5 – 6 MeV, the cross-section practically does not depend on the neutron energy, because the neutrons can reach any place in the nucleus with the same probability and so the ratio between the acts of fission and formation of U_{92}^{236} remains constant. For neutron energies above 6 MeV, the cross-section begins to increase, because of the "mechanical" fission.

Nuclear fission by a fast neutron is nuclear spitting caused by an energy which is the sum of the oscillation energy due to only one extra neutron and supplementary energy brought by the bombarding fast neutron. All nuclei with Z ≥ 90 undergo fission by fast neutrons. A low asymmetry (1 neutron in excess) is unable to produce a destructive force strong enough to split the nucleus. For a nuclear fission to occur it is necessary that the neutron from outside gets into the neutron-rich areas of the central zones of both branches. But because of the symmetry of even/even nuclei, the bonds in these zones are stronger. There are not enough large gaps between nucleons. The role of the energy is to help the neutron penetration and thus to form neutron asymmetry and to help the nuclear fission. For instance the fission of the nuclide U_{92}^{138} becomes possible only at neutron energy above 1 MeV. The speed of reaction becomes noticeable at neutron energy above 1,5 MeV. At energy of 2 MeV the cross-section of the fission becomes approximately 0,5. $10^{-24} cm^2$ [3] and remains constant at neutron energy increasing up to 6 MeV, which testifies. Hence the energy of this range provides equal probability of neutrons penetration in any place of the nucleus. The cross-section begins to increase at neutron energies above 5,5 - 6 MeV, because of the increasing speed of the "mechanical" fission.

The role of the transitive zone in nuclear fission with fast neutrons is relatively smaller. At Pu_{94}^{239} fission by fast neutrons the average number of prompt neutrons is 2,7 , i.e., near 3. This means, that fast neutrons penetrate mainly in the **pnn** groups, which form the central zone. Except that the binding energy in the nuclear center is smaller (see part 4.2). This predetermines that there is less difference between the mass of the fragments in fission by fast neutrons.

4.4.2. Remarks on nuclear fusion

The reactions of nuclear synthesis are known as thermonuclear fusion because of the considerable energy needed. But in any case, the energy released in the reaction is much more than the energy spent for inducing the reaction. Usually the electrostatic repulsion is considered as main obstacle which limits the speed of reactions. Of course, there is repulsion between the positive fields of the nuclei, but for the nuclear reactions the way of collision is also important. For instance, in reaction: $d + d = He_2^4$ + 24 MeV, in spite of the possibility for tremendous energy release, the reaction practically does not occur. Schematically the reaction can be represented as: ↑⇑ + ⇓↓=↑⇓⇑↓. A sideways collision of two deuterons with opposite spins is necessary. Because the two neutrons form a shield, the proton repulsion is of no significance. But the chance of sideways collision is very small because at motion the plane of the particle is parallel to the direction of motion. In addition, the collision must be strong enough to cause the spin exchange between neutrons and weak enough to avoid the scattering.

The reaction: $p + p = d + e^+ + v$ has a very small cross-section (10^{-23} barns) because it needs sideways collision of two protons having parallel spins and, in the short time of the collision, one of the protons must be converted into a neutron. The energy of collision must overcome the energy of repulsion, must keep together both protons until the proton stress causes "proton decay". It is assumed that this reaction takes place inside the stars, but it is very doubtful.

The small cross-section of the reactions: $p + d = He^3 + \gamma$ and $p + t = He^4 + \gamma$ is also predetermined by the need of sideways collisions. But the reaction: $d + t + He^4 + n + \gamma$ has a comparatively high cross-section (5 barns), because the separation of the neutron from the triton needs peripheral collisions which is more conventional.

The study of reactions of fusion between light nuclides is prompted by the interest in energy utilization. If the countless difficulties could be successfully overcome a practically inexhaustible source of energy will be found.

Synthesis of supper heavy nuclei. In principle the reactions of fusion are possible between any pair of nuclides. For instance, the reaction of synthesis of supper heavy mass nuclides is a type of reaction of fusion. The goal of these reactions is the discovery of eventual "island of stability". The probability of such "island" existence is very small because the nuclear stability depends on two main criteria.

1. *Magic number criterion.* As a matter of fact, the elements from Thorium (whose proton number 90 is a superposition of two magic numbers 82 + 8), to Californium (Z =98, or 82 + 16) present an island of stability after the gap of the very unstable elements after Bismuth. The superposition of the next magic numbers : 82 + 20 = 102 – is the element Nobelium, if it is an island, it is already discovered. The next "magic" combination: 82 + 50 = 132 is too far from the "shore".

2. *Proton stress criterion.* As it was shown above (see Fig 17) the proton stress in stable nuclides with maximum number of neutrons gradually increases, and at Z = 92 – 95 it will be equal to the proton stress in stable nuclides with minimum number of neutrons. At proton numbers above 95 the neutron shield becomes useless. At proton numbers 115 – 120, where the new

"island" of stability is assumed to be, the proton stress in nucleus will be commensurable with the proton stress between two protons.

Anyway, the synthesis of supper heavy mass nuclei will continue and the experiments are able to bring surprises.

CONCLUSION

The history of scientific development shows that every significant idea has at first been rejected. The more radical the idea, the more vigorous the rejection. The objective reasons for a rejection could be a natural mistrust or lack of convincing arguments. That is why the radical new idea about the nuclear structure is presented in this book in more detail, with numerous examples and explanations. It would be very difficult to reject the arguments because:

1. This is the only possible nuclear structure which is in agreement with *all* laws of physics. *This structure has no alternatives.*

2. All nuclear properties follow in a natural way from the proposed nuclear structure. As a supporting argument for this statement it would be enough to list several problem-puzzles: The strength of the He_2^4 nuclide; Mixed $\beta^+ and \beta^-$ decay; Nuclear fission by slow neutrons; The number of neutrons released in nuclear fission; Physical sense of the nuclear spin; The physical sense of magic numbers; Nature of nuclear isomerism etc.

3. The puzzled of the nuclear structure has been solved with the aid of the idea about the wave structure of the nucleons. This idea concerns the fundamental problem of the matter existence and is in agreement with the physical laws. But for the nuclear structure approval the acceptance of the idea about the wavy structure of matter is not necessary. It will be enough to admit the inevitable need of disk-like form of nucleons.

The subjective causes for rejection could be countless. The only flicker of hope is in the progress of humankind – we live in times different from Maxwell's…

REFERENCES

1. Nemetz O.F, Hofman U.V., Reference book on Nuclear Physics, Kiev, 1975

2. Encyclopedia of Physics, Moscow, 1987

3. Encyclopedic reference book of Physics, Moscow, 1984

4. Javorski V.M, Detlaf A.A, Reference book of Physics, Moscow, 1968

5. Glasston S.E, Sourcebook on Atomic Energy, New York, 1958

6. Key topics in Nuclear Structure – proceedings of the 8-th Int. Spring Seminar on Nuclear Physics, 23- 27 May 2004, Puestum, Italy

7. Slavov B. Introduction in theoretical Nuclear Physics, Sofia, 2003 (in Bulgarian)

8. Balabanov N, Nuclear Physics, Plovdiv, 1998 (in Bulgarian)

9. Evans E, Tritium and its compounds, Moscow, 1970

10. Halliday D, Resnic R, Fundamentals of Physics, New York, 1988

11. Feynman R, QED the strange theory of light and matter, Princeton, NJ, 1985.

12. Debye P, Ann. Physik, 39, 798, 1912.

13. Podkletnov E, Niminen R, Physica, C, Vol. 203, p. 442, (1992).

14. Kolessin R, About the Contrarevolution in Physics, Sofia, 2003, (in Bulgarian)

15. Gnedenko B.V, The Theory of Probability, Moscow, 1976, (in English)

16. Kolessin R., The Secret of Creation, Sofia, 1995, (in Bulgarian)